Principles of Bioinorganic Chemistry

Principles of Bioinorganic Chemistry

Freya Williams

MURPHY & MOORE
www.murphy-moorepublishing.com

Principles of Bioinorganic Chemistry
Freya Williams
ISBN: 978-1-63987-455-2 (Hardback)

MURPHY & MOORE
Published by Murphy & Moore Publishing,
1 Rockefeller Plaza,
New York City, NY 10020, USA

Cataloging-in-Publication Data

Principles of bioinorganic chemistry / Freya Williams.
p. cm.
Includes bibliographical references and index.
ISBN 978-1-63987-455-2
1. Bioinorganic chemistry. 2. Chemistry, Inorganic. 3. Biochemistry. I. Williams, Freya.
QP531 .P75 2021
572.51--dc23

Table of Contents

Preface

The field of chemistry which analyzes the role of metals in biology is known as bioinorganic chemistry. It draws on the principles of biochemistry and inorganic chemistry for studying the effects of substrate bindings and activation, electron-transfer proteins, group transfer chemistry, etc. Bioinorganic chemistry attempts to understand different phenomenon like the behavior of metal proteins, artificially induced metals and non-essential metals used in the fields of medicine and toxicology. This encompasses elements like iron, cobalt, zinc, copper, nickel, etc. Some of the important areas of research under this discipline are enzymology, toxicity, metal ion transport and storage, biomineralization, nitrogen fixation, and photosynthesis. This book attempts to understand the multiple branches that fall under the discipline of bioinorganic chemistry and how such concepts have practical applications. The topics included herein are of utmost significance and bound to provide incredible insights to readers. This will serve as a valuable source of reference for those interested in this field.

A short introduction to every chapter is written below to provide an overview of the content of the book:

Chapter 1 - The field of enquiry which focuses on the structures and biological functions of inorganic biological substances is known as bioinorganic chemistry. It studies the role of metals in biology. This is an introductory chapter which will introduce briefly all the significant aspects of bioinorganic chemistry as well as the inorganic compounds which are essential to human functioning; **Chapter 2** - There are various metal ions which play an integral role in biological systems. A few such metal ions are calcium, magnesium, sodium, cobalt, molybdenum, chromium, copper and iron ions. This chapter has been carefully written to provide an easy understanding of the varied facets of these metal ions in biological systems; **Chapter 3** - The proteins which have a metal ion as a cofactor are called metalloproteins. Most of the proteins fall in this category. A few types of metalloproteins are zinc enzymes and iron sulphur proteins. The topics elaborated in this chapter will help in gaining a better perspective about these metalloproteins as well as design of functional metalloproteins; **Chapter 4** - The process through which mineral crystals are produced by living organisms is known as biomineralization. It involves nucleation and crystal growth. Some of the minerals which are produced through this process are calcium phosphate and calcium carbonate. This chapter discusses in detail the process of biomineralization and these minerals; **Chapter 5** - The physical methods which are used within the field of bioinorganic chemistry can be broadly categorized into X-ray methods, magnetic resonance methods, Mössbauer spectroscopy and vibrational spectroscopy. This chapter closely examines these physical methods as well as their usage in bioinorganic chemistry to provide an extensive understanding of the subject; **Chapter 6** - There are various aspects which are studied within the domain of bioinorganic chemistry.

A few of them are biological nitrogen fixation, molybdenum nitrogenase, bioenergetics and glucose storage. The topics elaborated in this chapter will help in gaining a better perspective about these aspects of bioinorganic chemistry.

Finally, I would like to thank my fellow scholars who gave constructive feedback and my family members who supported me at every step.

Freya Williams

1

An Introduction to Bioinorganic Chemistry

The field of enquiry which focuses on the structures and biological functions of inorganic biological substances is known as bioinorganic chemistry. It studies the role of metals in biology. This is an introductory chapter which will introduce briefly all the significant aspects of bioinorganic chemistry as well as the inorganic compounds which are essential to human functioning.

Bioinorganic chemistry is the study of the structures and biological functions of inorganic biological substances, that is, those not containing carbon, such as metals. Bioinorganic chemistry is a field that examines the role of metals in biology. Bioinorganic chemistry includes the study of both natural phenomena such as the behavior of metalloproteins as well as artificially introduced metals, including those that are non-essential, in medicine and toxicology. There is a particular interest in this field when one is interested to understand the chemistry behind biological processes.

As a combination of biochemistry and inorganic chemistry, bioinorganic chemistry is important in elucidating the implications of electron-transfer proteins, substrate bindings and activation, atom and group transfer chemistry as well as metal properties in biological chemistry. The discipline also includes the study of inorganic models or mimics that imitate the behavior of metalloproteins. It is important not only for chemists but even for biologists to understand the mechanism of how biological systems work.

Many biological reactions are known to involve metal ions. The following list shows typical bioactive substances containing metals:

- Electron carriers. Fe: Cytochrome, iron-sulfur protein. Cu: Blue copper protein.
- Metal storage compound. Fe: Ferritin, transferrin. Zn: Metallothionein.
- Oxygen transportation agent. Fe: Hemoglobin, myoglobin. Cu: Hemocyanin.
- Photosynthesis. Mg: Chlorophyll.
- Hydrolase. Zn: Carboxypeptidase. Mg: Aminopeptidase.
- Oxidoreductase. Fe: Oxygenase, hydrogenase. Fe, Mo: Nitrogenase.
- Isomerase. Fe: Aconitase. Co: Vitamin B_{12} coenzyme.

The basis of chemical reactions of metalloenzymes are:

- Coordinative activation (coordination form, electronic donating, steric effect).
- Redox (metal oxidation state).
- Information communication, and, in many cases, reaction environments are regulated by biopolymers such as proteins, and selective reactions are performed.

Examples of actions of metals other than by metalloenzymes include:

- Mg: MgATP energy transfer.
- Na/K ion pumping.
- Ca: Transfer of hormone functions, muscle contraction, nerve transfer, blood coagulation, are some of the important roles of metals.

Oxidation

Oxidation reactions in living systems are fundamental to life, and many studies of these systems have been performed. In particular, the mechanisms of oxygen gas transportation by hemoglobin and mono-oxygen oxidation by the iron porphyrin compounds named P-450 have been studied at length. Oxygen gas transportation, which has been studied for many years, is described below. Iron porphyrins hemoglobin and myoglobin and the copper compound hemocyanin are involved in the transportation of oxygen gas in air to cells in living organisms. The basis of this function is reversible bonding and dissociation of dioxygen to iron or copper ions. In order to perform these functions, metals must be in oxidation states and coordination environments suitable for the reversible coordination of dioxygen. The iron porphyrin compound hemoglobin is found in red bloods of human beings and other animals.

Hemoglobin has the structure of heme iron with four iron porphyrin units combined with a globin protein. Dioxygen is transported in blood by being coordinated to ferrous ions in the hem iron unit. The Fe (II) ion is penta-coordinate with four nitrogen atoms of porphyrin and a nitrogen atom of the polypeptide histidine, and becomes hexa-coordinate when a dioxygen coordinates to it. The spin state of Fe (II) changes from high spin to low spin upon the coordination of dioxygen. The high spin Fe(II) is above the plane of porphyrin because it is too large to fit in the available space. When the Fe(II) ion becomes low spin upon dioxygen coordination, the size of the iron ion decreases and it just fits into the hole of the porphyrin molecule.

This molecular-level movement has attracted interest as an allosteric effect because it affects the whole protein through the histidine coordinate bond and governs the specific bond of a dioxygen molecule. Oxidation of the Fe(II) ion of a hem molecule is prevented by a macromolecular protein, and if the hem iron is taken out of the prote in, Fe(II) ion is oxidized to Fe(III), and two porphyrin rings are bridged by a peroxideμ-O_2^{2-}, which finally changes to a bridgingμ-O_2-structure.

When the hem is in this state, it loses the ability to coordinate to the dioxygen molecule. Based on this phenomenon, a synthetic porphyrin that is able reversibly to coordinate to a dioxygen by suppressing dimerization of the iron porphyrin has been developed, and was named the picket fence porphyrin after its three dimensional form.

Nitrogen Fixation

The reaction which converts the nitrogen in air into ammonia is basic to all life. Nitrogen fixation, the reaction to fix atmospheric nitrogen to form ammonia, is carried out by Rhizobium in the roots of legumes or by bacteria in algae in an anaerobic atmosphere. All animals and plants, including mankind, were depended on biological nitrogen fixation as a source of nitrogen for protein and other compounds containing nitrogen before the invention of the Harber-Bosch process.

$$N_2 + 8H^+ + 8e^- + 16MgATP \rightarrow 2NH_3 + H_2 + 16MgADP + 16Pi$$

(where Pi is an inorganic phosphate)

An enzyme named nitrogenase catalyzes this reaction. Nitrogenase contains iron-sulfur and iron-molybdenum sulfur proteins, and reduces dinitrogen by coordination and cooperative proton and electron transfers, while using MgATP as an energy source. Because of the importance of this reaction, attempts to clarify the structure of nitrogenase and to develop artificial catalysts for nitrogen fixation have continued for many years. Recently, the structure of an active center in nitrogenase called iron-molybdenum cofactor was clarified by single crystal X-ray analysis. According to this analysis, its structure has Fe_3MoS_4 and Fe_4S_4 clusters connected through S. It is believed that dinitrogen is activated by coordination between the two clusters. On the other hand, the portion called P cluster consists of two Fe_4S_4 clusters. The roles and reaction mechanism of both parts are not yet clear.

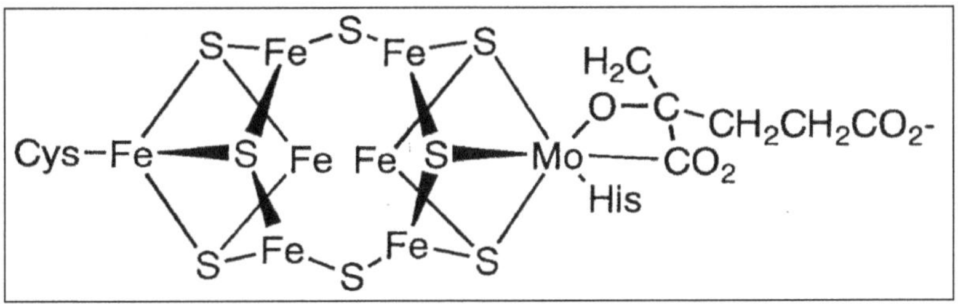

Structure of Fe-Mo cofactor in nitrogenase.

Photosynthesis

The formation of glucose and dioxygen by the reaction of carbon dioxide and water is a skillful reaction using photo energy and in which chlorophyll, which is a magnesium porphyrin and a manganese cluster complex, plays the central role. A chloroplast contains photosystem I (PSI) and photosystem II (PSII), which use light energy to reduce carbon dioxide and to oxidize water.

Chlorophyll is a fundamental component of PSI. Chlorophyll is a porphyrin complex of magnesium and is responsible for the green colors of leaves. It plays an important role in receiving light energy and transferring it to redox reaction systems. Chlorophyll is excited from the singlet ground state to the singlet excited state by light, the energy of the excited state is transferred to an acceptor within 10 ps, and the resultant energy reduces an iron-sulfur complex and is finally used for reduction of carbon dioxide in subsequent dark reactions. Since charge separation by photochemical excitation is the most important first stage, studies on photo induced electron transfer are have been actively performed using various kinds of porphyrin compounds as models of chlorophylls. PSI, which obtains oxidizing energy by electron transfer, converts ADP to ATP.

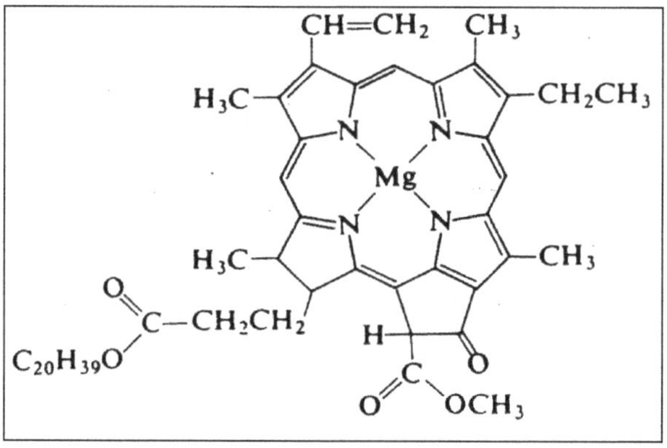

Chlorophyll a.

On the other hand, the oxidized form of PSII oxidizes water through a chain of redox reactions of oxo cluster complexes of manganese, and generates oxygen. Since four electrons shift in the reduction of Mn(IV) to Mn (II) in this reaction, at least two manganese species are involved. Probably, a cluster complex which contains two Mn(II) and two Mn(IV) species mediates the electron transfer via four step reactions. However, the details of this reaction are as yet unclear because it is very difficult to isolate this cluster and to analyze its structure. The electron transfer stage is being studied at present by using various manganese complexes as model systems.

Photosynthesis is a very interesting research theme in bioinorganic chemistry as it involves a few metal ions, a porphyrin, sulfide and oxide clusters that constitute a cycle of subtle electron transfer and redox reactions, and generates oxygen gas by photolysis of water and produce carbohydrates from carbon dioxide by reductive dark reactions.

Inorganic Compounds Essential to Human Functioning

Inorganic compounds are important in the body and responsible for many simple functions. The major inorganic compounds are water (H_2O), bimolecular oxygen (O_2), carbon dioxide (CO_2), and some acids, bases, and salts. The body is composed of 60–75% water. Oxygen is required by all cells for cellular metabolism and circulating blood must be well oxygenated for maintenance of life. Carbon dioxide is a waste product of cells and must be eliminated or a serious change in pH can occur, known as acidosis. A balance in acids, bases, and salts must be maintained to assure homeostasis of blood pH and electrolyte balance.

As much as 70 percent of an adult's body weight is water. This water is contained both within the cells and between the cells that make up tissues and organs. Its several roles make water indispensable to human functioning.

Water as a Lubricant and Cushion

Water is a major component of many of the body's lubricating fluids. Just as oil lubricates the hinge on a door, water in synovial fluid lubricates the actions of body joints, and water in pleural fluid helps the lungs expand and recoil with breathing. Watery fluids help keep food flowing through the digestive tract, and ensure that the movement of adjacent abdominal organs is friction free. Water also protects cells and organs from physical trauma, cushioning the brain within the skull, for example, and protecting the delicate nerve tissue of the eyes. Water cushions a developing fetus in the mother's womb as well.

Water as a Heat Sink

A heat sink is a substance or object that absorbs and dissipates heat but does not experience a corresponding increase in temperature. In the body, water absorbs the heat generated by chemical reactions without greatly increasing in temperature. Moreover, when the environmental temperature soars, the water stored in the body helps keep the body cool. This cooling effect happens as warm blood from the body's core flows to the blood vessels just under the skin and is transferred to the environment. At the same time, sweat glands release warm water in sweat. As the water evaporates into the air, it carries away heat, and then the cooler blood from the periphery circulates back to the body core.

Water as a Component of Liquid Mixtures

A mixture is a combination of two or more substances, each of which maintains its own chemical identity. In other words, the constituent substances are not chemically bonded into a new, larger chemical compound. The concept is easy to imagine if you think of powdery substances such as flour and sugar; when you stir them together in a bowl, they obviously do not bond to form a new compound. The room air you breathe is a gaseous mixture, containing three discrete elements—nitrogen, oxygen, and argon—and one compound, carbon dioxide. There are three types of liquid mixtures, all of which contain water as a key component; these are solutions, colloids, and suspensions.

For cells in the body to survive, they must be kept moist in a water-based liquid called a solution. In chemistry, a liquid solution consists of a solvent that dissolves a substance called a solute. An important characteristic of solutions is that they are homogeneous; that is, the solute molecules are distributed evenly throughout the solution. If you were to stir a teaspoon of sugar into a glass of water, the sugar would dissolve into sugar molecules separated by water molecules. The ratio of sugar to water in the left side of the glass would be the same as the ratio of sugar to water in the right side of the glass. If you were to add more sugar, the ratio of sugar to water would change, but the distribution—provided you had stirred well—would still be even.

Water is considered the "universal solvent" and it is believed that life cannot exist without water because of this. Water is certainly the most abundant solvent in the body; essentially all of the body's chemical reactions occur among compounds dissolved in water. Since water molecules are polar, with regions of positive and negative electrical charge, water readily dissolves ionic compounds and polar covalent compounds. Such compounds are referred to as hydrophilic, or "water-loving." As mentioned above, sugar dissolves well in water. This is because sugar molecules contain regions of hydrogen-oxygen polar bonds, making it hydrophilic. Nonpolar molecules, which do not readily dissolve in water, are called hydrophobic, or "water-fearing."

Concentrations of Solutes

The concentration of a given solute is the number of particles of that solute in a given space (oxygen makes up about 21 percent of atmospheric air). In the bloodstream of humans, glucose concentration is usually measured in milligram (mg) per deciliter (dL), and in a healthy adult averages about 100 mg/dL. Another method of measuring the concentration of a solute is by its molarilty—which is moles (M) of the molecules per liter (L). The mole of an element is its atomic weight, while a mole of a compound is the sum of the atomic weights of its components, called the molecular weight. An often-used example is calculating a mole of glucose, with the chemical formula $C_6H_{12}O_6$. Using the periodic table, the atomic weight of carbon (C) is 12.011 grams (g), and there are six carbons in glucose, for a total atomic weight of 72.066 g. Doing the same calculations for hydrogen (H) and oxygen (O), the molecular weight equals 180.156g (the "gram molecular weight" of glucose). When water is added to make one liter of solution, you have one mole (1M) of glucose. This is particularly useful in chemistry because of the relationship of moles to "Avogadro's number." A mole of any solution has the same number of particles in it: 6.02×10^{23}. Many substances in the bloodstream and other tissue of the body are measured in thousandths of a mole, or millimoles (mM).

A colloid is a mixture that is somewhat like a heavy solution. The solute particles consist of tiny clumps of molecules large enough to make the liquid mixture opaque (because the particles are large enough to scatter light). Familiar examples of colloids are milk and cream. In the thyroid glands, the thyroid hormone is stored as a thick protein mixture also called a colloid.

A suspension is a liquid mixture in which a heavier substance is suspended temporarily in a liquid, but over time, settles out. This separation of particles from a suspension is called sedimentation. An example of sedimentation occurs in the blood test that establishes sedimentation rate, or sed rate. The test measures how quickly red blood cells in a test tube settle out of the watery portion of blood (known as plasma) over a set period of time. Rapid sedimentation of blood cells does not normally happen in the healthy body, but aspects of certain diseases can cause blood cells to clump together, and these heavy clumps of blood cells settle to the bottom of the test tube more quickly than do normal blood cells.

The Role of Water in Chemical Reactions

Two types of chemical reactions involve the creation or the consumption of water: Dehydration synthesis and Hydrolysis.

- In dehydration synthesis, one reactant gives up an atom of hydrogen and another reactant gives up a hydroxyl group (OH) in the synthesis of a new product. In the formation of their covalent bond, a molecule of water is released as a byproduct. This is also sometimes referred to as a condensation reaction.
- In hydrolysis, a molecule of water disrupts a compound, breaking its bonds. The water is itself split into H and OH. One portion of the severed compound then bonds with the hydrogen atom, and the other portion bonds with the hydroxyl group.

These reactions are reversible, and play an important role in the chemistry of organic compounds.

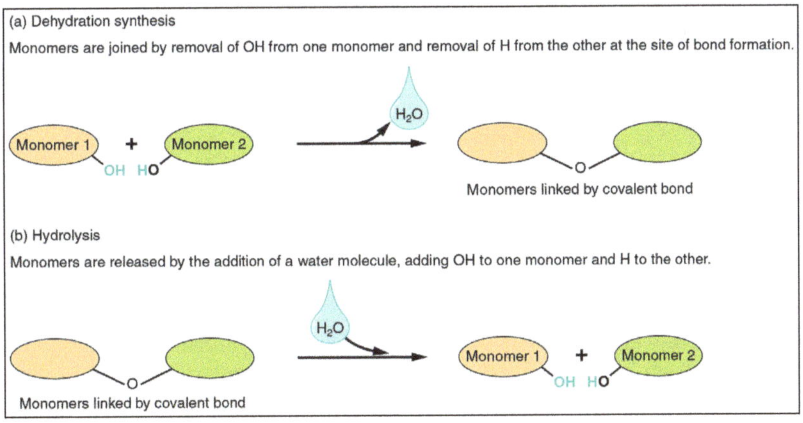

Dehydration Synthesis and Hydrolysis: Monomers, the basic units for building larger molecules, form polymers (two or more chemically-bonded monomers): (a) In dehydration synthesis, two monomers are covalently bonded in a reaction in which one gives up a hydroxyl group and the other a hydrogen atom. A molecule of water is released as a byproduct during dehydration reactions. (b) In hydrolysis, the covalent bond between two monomers is split by the addition of a hydrogen atom to one and a hydroxyl group to the other, which requires the contribution of one molecule of water.

Salts

Recall that salts are formed when ions form ionic bonds. In these reactions, one atom gives up one or more electrons, and thus becomes positively charged, whereas the other accepts one or more electrons and becomes negatively charged. You can now define a salt as a substance that, when dissolved in water, dissociates into ions other than H^+ or OH^-. This fact is important in distinguishing salts from acids and bases.

A typical salt, NaCl, dissociates completely in water. The positive and negative regions on the water molecule (the hydrogen and oxygen ends respectively) attract the negative chloride and positive sodium ions, pulling them away from each other. Again, whereas nonpolar and polar covalently bonded compounds break apart into molecules in solution, salts dissociate into ions. These ions are electrolytes; they are capable of conducting an electrical current in solution. This property is critical to the function of ions in transmitting nerve impulses and prompting muscle contraction.

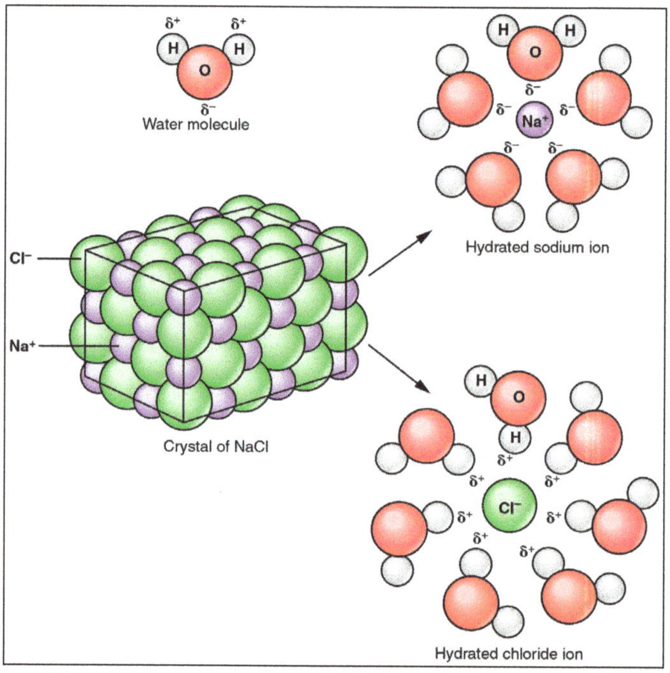

Dissociation of Sodium Chloride in Water: The crystals of sodium chloride dissociate not into molecules of NaCl, but into Na^+ cations and Cl^- anions, each completely surrounded by water molecules.

Many other salts are important in the body. For example, bile salts produced by the liver help break apart dietary fats, and calcium phosphate salts form the mineral portion of teeth and bones.

Acids and Bases

Acids and bases, like salts, dissociate in water into electrolytes. Acids and bases can very much change the properties of the solutions in which they are dissolved.

Acids

An acid is a substance that releases hydrogen ions (H^+) in solution. Because an atom of hydrogen has just one proton and one electron, a positively charged hydrogen ion is simply a proton. This solitary proton is highly likely to participate in chemical reactions. Strong acids are compounds that release all of their H^+ in solution; that is, they ionize completely. Hydrochloric acid (HCl), which is released from cells in the lining of the stomach, is a strong acid because it releases all of its H^+ in the stomach's watery environment. This strong acid aids in digestion and kills ingested microbes. Weak acids do not ionize completely; that is, some of their hydrogen ions remain bonded within a compound in solution. An example of a weak acid is vinegar, or acetic acid; it is called acetate after it gives up a proton.

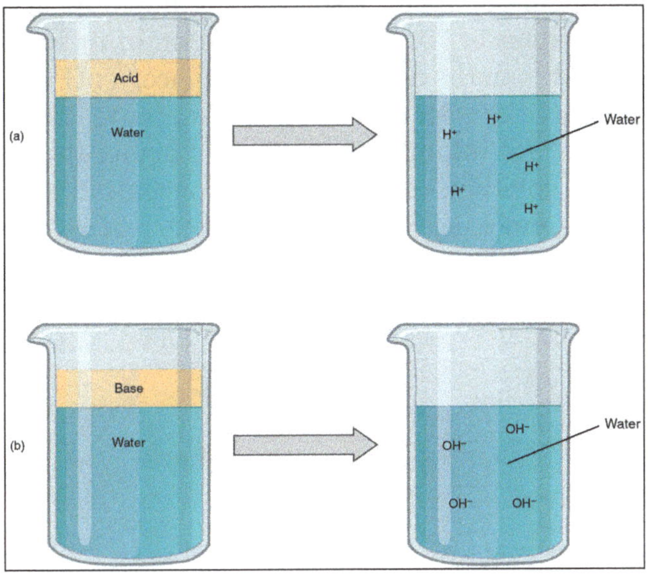

Acids and Bases: (a) In aqueous solution, an acid dissociates into hydrogen ions (H^+) and anions. Nearly every molecule of a strong acid dissociates, producing a high concentration of H^+. (b) In aqueous solution, a base dissociates into hydroxyl ions (OH^-) and cations. Nearly every molecule of a strong base dissociates, producing a high concentration of OH^-.

Bases

A base is a substance that releases hydroxyl ions (OH^-) in solution, or one that accepts H^+ already present in solution. The hydroxyl ions (also known as hydroxide ions) or other basic substances combine with H^+ present to form a water molecule, thereby removing H^+ and reducing the solution's acidity. Strong bases release most or all of their hydroxyl ions; weak bases release only some hydroxyl ions or absorb only a few H^+. Food mixed with hydrochloric acid from the stomach would burn the small intestine (the next portion of the digestive tract after the stomach), if it were not for the release of bicarbonate (HCO_3^-), a weak base that attracts H^+. Bicarbonate accepts some of the H^+ protons, thereby reducing the acidity of the solution.

The Concept of pH

The relative acidity or alkalinity of a solution can be indicated by its pH. A solution's pH is the negative, base-10 logarithm of the hydrogen ion (H^+) concentration of the solution. As an example, a pH 4 solution has an H^+ concentration that is ten times greater than that of a pH 5 solution. That is, a solution with a pH of 4 is ten times more acidic than a solution with a pH of 5. The concept of pH will begin to make more sense when you study the pH scale. The scale consists of a series of increments ranging from 0 to 14. A solution with a pH of 7 is considered neutral—neither acidic nor basic. Pure water has a pH of 7. The lower the number below 7, the more acidic the solution, or the greater the concentration of H^+. The concentration of hydrogen ions at each pH value is 10 times different than the next pH. For instance, a pH value of 4 corresponds to a proton concentration of 10^{-4} M, or 0.0001M, while a pH value of 5 corresponds to a proton concentration of 10^{-5} M, or 0.00001M. The higher the number above 7, the more basic (alkaline) the solution, or the lower the concentration of H^+. Human urine, for example, is ten times more acidic than pure water, and HCl is 10,000,000 times more acidic than water.

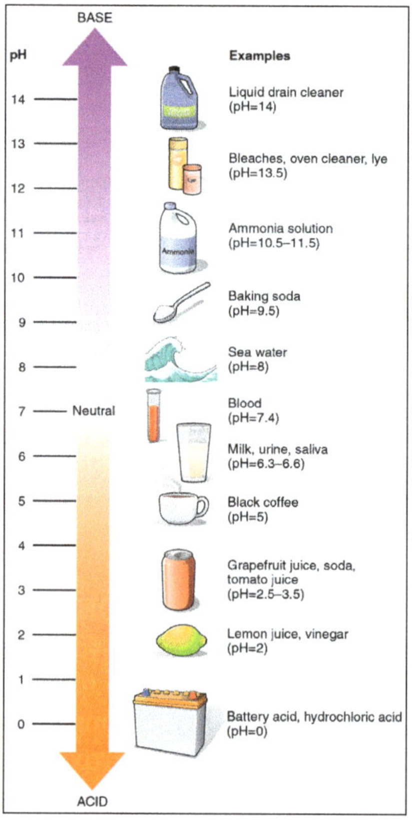

The pH Scale.

Buffers

The pH of human blood normally ranges from 7.35 to 7.45, although it is typically identified as pH 7.4. At this slightly basic pH, blood can reduce the acidity resulting from the carbon dioxide (CO_2) constantly being released into the bloodstream by the trillions of cells in the body. Homeostatic mechanisms (along with exhaling CO_2 while breathing) normally keep the pH of blood within this narrow range. This is critical, because fluctuations—either too acidic or too alkaline—can lead to life-threatening disorders.

All cells of the body depend on homeostatic regulation of acid–base balance at a pH of approximately 7.4. The body therefore has several mechanisms for this regulation, involving breathing, the excretion of chemicals in urine, and the internal release of chemicals collectively called buffers into body fluids. A buffer is a solution of a weak acid and its conjugate base. A buffer can neutralize small amounts of acids or bases in body fluids. For example, if there is even a slight decrease below 7.35 in the pH of a bodily fluid, the buffer in the fluid—in this case, acting as a weak base—will bind the excess hydrogen ions. In contrast, if pH rises above 7.45, the buffer will act as a weak acid and contribute hydrogen ions.

Homeostatic Imbalance

The excessive acidity of acids and bases, of the blood, and other body fluids is known as acidosis. Common causes of acidosis are situations and disorders that reduce the effectiveness of breathing, especially the person's ability to exhale fully, which causes a buildup of CO_2 (and H^+) in the bloodstream. Acidosis can also be caused by metabolic problems that reduce the level or function of buffers that act as bases, or that promote the production of acids. For instance, with severe diarrhea, too much bicarbonate can be lost from the body, allowing acids to build up in body fluids. In people with poorly managed diabetes (ineffective regulation of blood sugar), acids called ketones are produced as a form of body fuel. These can build up in the blood, causing a serious condition called diabetic ketoacidosis. Kidney failure, liver failure, heart failure, cancer, and other disorders also can prompt metabolic acidosis.

In contrast, alkalosis is a condition in which the blood and other body fluids are too alkaline (basic). As with acidosis, respiratory disorders are a major cause; however, in respiratory alkalosis, carbon dioxide levels fall too low. Lung disease, aspirin overdose, shock, and ordinary anxiety can cause respiratory alkalosis, which reduces the normal concentration of H^+.

Metabolic alkalosis often results from prolonged, severe vomiting, which causes a loss of hydrogen and chloride ions (as components of HCl). Medications can also prompt alkalosis. These include diuretics that cause the body to lose potassium ions, as well as antacids when taken in excessive amounts, for instance by someone with persistent heartburn or an ulcer.

References

- BIOINORGANIC-CHEMISTRY-1465064196: iitkgp.ac.in, Retrieved 18, February 2020
- A-Reaction-and-Physical-Properties, A-Inorganic-Chemistry-(Saito): libretexts.org, Retrieved 25, July 2020
- Inorganic-compounds-essential-to-human-functioning: oregonstate.education, Retrieved 02, March 2020

2

Metal Ions in Biological Systems

There are various metal ions which play an integral role in biological systems. A few such metal ions are calcium, magnesium, sodium, cobalt, molybdenum, chromium, copper and iron ions. This chapter has been carefully written to provide an easy understanding of the varied facets of these metal ions in biological systems.

Metals play crucial roles in life processes. It is increasingly recognized that metals are involved in cellular and subcellular functions. With the application of new and sophisticated machines to study biological and biochemical systems the true role of inorganic salts in living systems can be revealed. Inorganic chemistry is not the "Dead Chemistry" that some people may think. Today, it is known that metals are important ingredients in life, just as the organic molecules. For instance, the divalent magnesium and calcium ions play important regulatory roles in cells. Metallothionein are proteins rich in metal ions found in living systems. The divalent cations Zn^{2+}, Ca^{2+} and Mg^{+} prevent cytotoxicity and in vivo antagonize Cd-induced carcinogenesis. Lack of body iron is common in cancer patients and it is associated with complications in surgery and in animal experiments. The transport of iron and other metal ions by the blood plasma is achieved through the formation of protein complexes. Copper is recognized as an essential metalloelement and is primarily associated with copper-dependent cellular enzymes. Metals are also used as inorganic drugs for many diseases.

Elements in Our Body

Consider the content of the elements in the body of an average healthy person (weighing 70 kg). It has been established that more than half of the total weight of an average man is accounted by oxygen. The amount of various elements in our body is given in table.

Table: Average concentration of various elements present in our body.

Elements	Amount in our Body
O	45.5 kg
C	12.6kg
H	7.0 kg
N	2.10 kg
P	700 g
Ca	1050 g

K	140 g
Na	105 g
Mg	35 g
Fe	4.2 g
Zn	2.3 g

The content of the rest of metals does not exceed one gram particularly Cu –0.11 g and Mn – 0.02 g. So only 2% of the human body is composed of metals, but much of the life depends on them.

Geochemical Effects on the Distribution of Metals

All the essential elements, except Mo are fairly abundant in the earth's crust. Although Al (8.2%), Si (28.2%), Ti (0.57%) and Zr (0.02%) are abundant, these are not essential elements. Since all these elements form insoluble oxides at biological pH and do not form stable complexes with complexing agents of biological significance, they are non-toxic at normal levels. Though they can be harmful at very high levels. All the well-known toxic elements are extremely rare in the earth's crust: As (~2 x 10^{-4} %), Pb (~ 1.3 x 10^{-3} %), Cd (~2 x 10^{-5} %) and Hg (~5 x 10^{-5} %).

Classification of Elements

Essential Elements and Trace Elements

Of the 109 known elements, 30 elements are believed to be essential for the survival of living organisms. Nineteen of the 30 are trace elements, out of which 12 are transition metals. We can account for the biological activity of about one-half of the essential trace elements which function in metalloenzymes, including Fe, Zn, Cu, Mn, Mo, Co, Ni and Se.

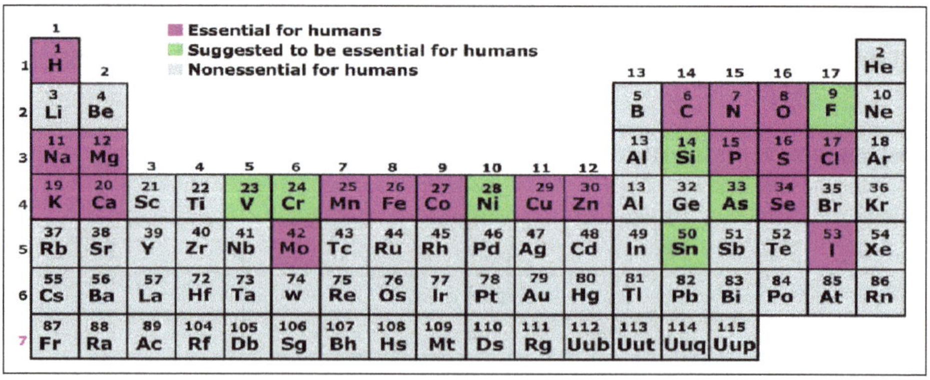

Essential elements shown in the periodic table.

Knowledge of the biological function of the trace elements has lagged far behind. The treatment of anemia with iron and the association of iodine deficiency with goiter marked these as the only two trace elements recognized as essential for animals. In an early 20th century discovery of the essentiality of copper, zinc, cobalt, manganese, and molybdenum in animals were the main thrust. These efforts have resulted in evidence supporting the essentiality of selenium, chromium, tin, vanadium, fluorine, silicon, nickel, lead, cadmium, arsenic, and lithium. The criteria for an element to play an essential role in our body are as follows:

- A physiological deficiency appears when the element is removed from a purified diet.
- The deficiency can be relieved by the addition of one specific element.
- A specific biochemical function is associated with a particular element.

The simplest definition of an essential element is that, it is an element required for the maintenance of life; its absence results in death or a severe malfunction of the organism. This rigorous criterion cannot always be satisfied and this has led to a broader definition of essentiality. An element is considered essential when a deficient intake produces an impairment of function and when restoration of physiological levels of that element relieves the impaired function or prevents impairment. The organism can neither grow nor complete its life cycle without the element in question. The element should have a direct influence on the organism and be involved in its metabolism. The effect of the essential element cannot be wholly replaced by any other element. The essential trace elements provide a classic example of required nutrients. An organism passes through several stages as the concentration of essential nutrient progresses from deficiency to excess. In absolute deficiency, death may occur.

Essential Ultratrace Elements

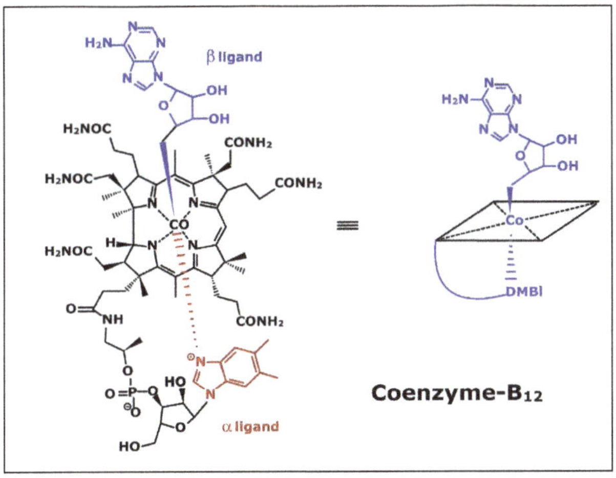

Co ion in coenzyme B_{12}.

The essential ultratrace elements are required for growth and survival of cells and organisms. They normally occur and function in cells at extremely low concentrations, usually far less than 1μM and as low as 10^{-8} to 10^{-9} M.

Of the essential ultratrace metals only four (Mn, Mo, Co, Ni) have been clearly identified as forming metallo-enzymes. Manganese metalloenzymes are present for several important enzymes superoxide dismutase (mitochondrial), arginase, pyruvate carboxylase, and glycosyl transferase. In molybdo-metalloenzymes, the metal exists as a Mo cofactor, a complex of the metal with novel organic molecules, called molybdopterin. The role of cobalt is the best understood of any of the essential ultratrace metals. Both vitamin B_{12} and coenzyme B_{12}, have a cobalt ion complexed in their equatorial positions by four nitrogens of a macrocyclic ligand called corrin.

Essential Ultratrace Nonmetals

The essential non-metals comprise a much more heterogeneous group then do the transition metals, whose properties are all closely related. The two essential halogens, F and I are highly specific and very different. Fluoride has a remarkable anti-dental caries effect. This may be related to its ability to replace OH, thereby stabilizing the structural matrix of bones and teeth. Fluoride also can inhibit strongly certain key enzyme, enolase, pyrophosphatase. Among the non-metals the most important is selenium. Despite its high toxicity, selenium has been shown to be a component of several enzymes involved in essential oxidation-reduction reactions.

Toxic Elements

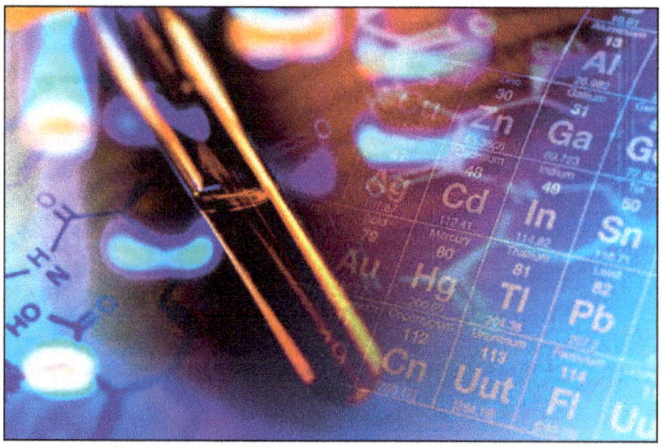

Toxic elements, like chemical toxins, are ubiquitous in your environment. Depending on where you live and where your foods (including herbs) are grown, your toxic element exposures can vary widely. In addition, agriculture practices (fungicides, herbicides, pesticides, etc.), consumer products (from personal products to building materials), and medicines (prescription and over-the-counter) increase your exposures to toxic elements. Because of their deliberate uses for decades, toxic elements are now in the air, water, and soil (food sources – plant and animal). As such, whether you realize it or not, you are continually exposed to a variety of toxic elements to some degree.

Toxic metals, including "heavy metals," are individual metals and metal compounds that negatively affect people's health. Some toxic, semi-metallic elements, including arsenic and selenium. In very small amounts, many of these metals are necessary to support life. However, in larger amounts, they become toxic. They may build up in biological systems and become a significant health hazard.

Toxicity

Mechanism of the toxicity of metals is very complicated. Generally toxicity of metals may result from one of the following:

- Blocking the essential biological functional groups of biomolecules such as enzymes. Amino acid residues like serine is –OH functional group, cysteine is –SH group and histidine –N group often constitute the active sites of enzymes. A toxic metal ion may bind with these functional groups and block the activity of the enzyme.
- Displacing the essential metal ions from biomolecules. A biomolecule with a foreign metal ion loses its activity.
- Modifying the active conformation of biomolecules. Biomolecules are having specific active conformations and if this active conformation is lost due to the coordination of a metal ion, the activity of the biomolecule is lost.

Toxicity and Deficiency

Most of the essential elements are also toxic when taken in excess. Too much copper causes necrotic hepatitis and hemolytic anemia. Selenium is the most toxic of the essential element, and intakes greater than 4 $\mu g\ g^{-1}$ is serious. Too much manganese causes psychic and neurological disorders. The dual action of trace elements, in relation to toxicity and deficiency is represented by a dose-response curve. The cut-off points, between deficiency, normal health and toxicity, vary from nutrient and nutrient, and from person to person. Both insufficient and an excess of iodine produces thyroid problems. Deficiency of Cu and Se may contribute to kwashiorkor. Deficiency in Co, a component of vitamin B_{12}, can lead to anaemia.

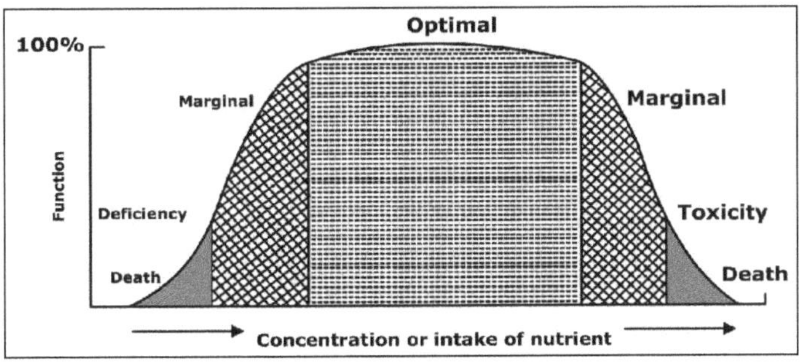

Dose response range of an essential element.

The nature of the chemical form of an element influences the dose response relationship. For example, the cations Sn^{2+} and Ni^{2+} are not particularly toxic, the chemical species $Sn(CH_3)_3^+$ and $Ni(CO)_4$ are dangerously toxic. In these cases the toxicity is due to the whole species, not just the trace element. In the case of $Ni(CO)_4$ the toxicity arises from the dissociation of the CO groups. Chromium (VI) is more toxic than chromium (III) which is due to its strong oxidising properties.

Toxic Effects of Trace Elements

Lead

Lead is so widespread in our society it is probably the most serious toxic metal. Around 70% to 90% of the lead assimilated goes into the bones, the next major accumulator is the liver, then the kidneys. Divalent lead readily replaces Ca^{2+} in bone, and either becomes firmly fixed or reversibly fixed. Lead in the latter form may be released into the blood stream and a person who has been purged of lead may have a reoccurrence of lead poisoning at a later date to release from bone tissue.

Symptoms of lead poisoning are many and not always the same for every person. The principal danger areas are, the hematopoietic system, leading to anaemia, the nervous system, leading to irreversible brain damage, and the renal system. Young children are particularly susceptible, the damage showing up in behavioral and educational abnormalities, and may be mental retardation.

Arsenic

Arsenic has similar toxic properties to Pb, Hg and Cd, as regards bonding to sulphur and inhibiting enzyme action such as pyruvate dehyrogenase. Arsenic, which is found mainly in the liver, kidneys, lungs and intestinal walls, is readily absorbed if water soluble. Since P and As are in the same periodic group, and since P is a major essential element, it is not surprising that arsenic interferes with phosphorous metabolism. Thus arsenic is toxic by replacement by phosphorous in ATP and in coagulation of proteins.

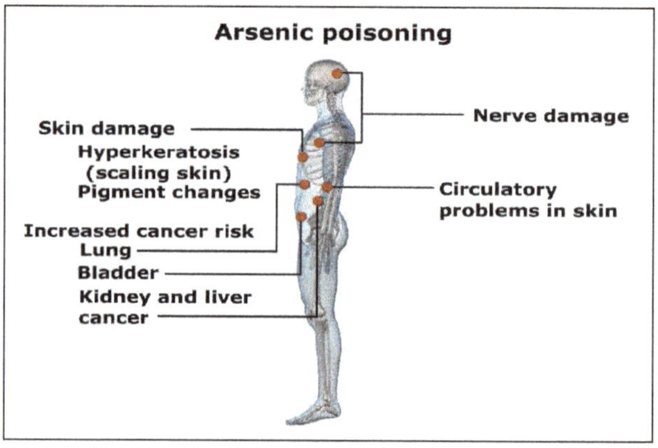

Arsenic poisoning.

Cadmium

Cadmium in lower levels has an adverse effect on human beings. Tobacco contains around 1 μg g^{-1} of Cd, so up to 2 to 4 μg of metal are inhaled from smoking one packet. Approx. 25 to 50% of inhaled cadmium is absorbed into the body, while from food the absorption is about 6%. A lethal dose of Cd is > 350 mg. Cadmium accumulates in the liver and kidneys produces irreversible kidney damage.

Respiratory and pulmonary damage is reported to occur from the breathing of cadmium vapour or particulates. Cadmium cannot cross the placental membrane and the mammary gland is an effective barrier so it does not affect the central nervous system. Therefore new-born babies are quite free of any cadmium. When excessive amounts of cadmium are ingested, it replaces zinc at key enzymatic sites, bringing about metabolic disorders.

Mercury

Mercury is the most toxic heavy metal, and many serious incidences have resulted from mercury poisoning. The toxicity of mercury is related to the chemical form. Liquid mercury appears to have little effect, but mercury vapour is readily absorbed into the bloodstream producing brain damage.

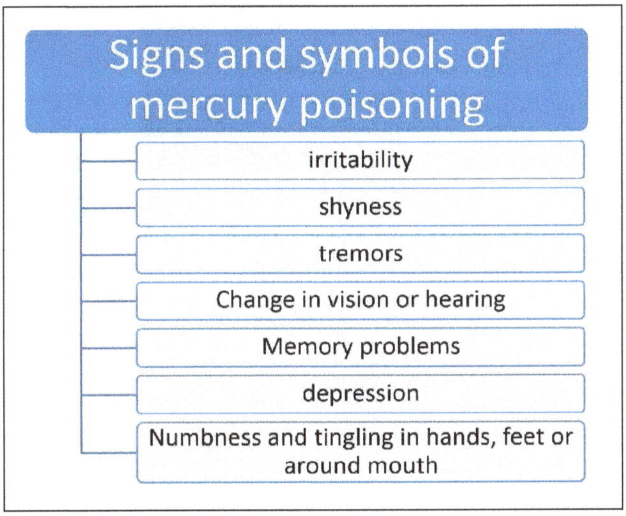

Symptoms of Mercury poisoning.

Mercury as a toxic metal came to the limelight after the incidence of "Minamata disease" in 1953-60 in Japan. Nearly 111 cases of mercury poisoning were reported among persons who had eaten mercury contaminated fish from Minamata Bay. Among them, about 45 people died. Genetic defects had been observed in 20 babies whose mothers had eaten seafood from the bay. Elemental mercury has been fairly inert and non-toxic. If it is swallowed, it gets excreted without causing serious damage. Hg vapour, when inhaled, enters the brain through the blood stream, giving rise to severe damage of the central nervous system. Alkyl mercurials are even more dangerous.

Copper

Copper is essential for all forms of life, problems arise when it is deficient or in excess. In some respects intake of essential elements is more critical than for toxic elements. The daily human consumption of Cu is around 2-5 mg, while the required amount for a child is 1 to 1.6 mg per day, and for adult 2mg per day while less than 0.3 mg per day produces deficiency. Copper is mainly distributed in the liver, kidneys and intestines.

Copper deficiency means reduced protein and enzyme activity. Signs of this are: anaemia, loss of hair pigment, reduced growth and reduced arterial elasticity. Menkes kinky hair disease in children is due to a deficiency of cytochrome oxidase, and is fatal by the age 3-5 years. Wilson's disease occurs with excess Cu where the biosynthesis of ceruloplasmin is suppressed, it is an inherited disease.

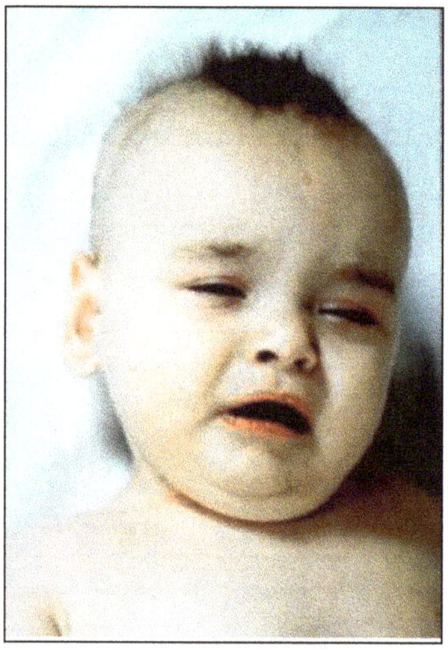

Menkes kinky hair disease in children.

Zinc

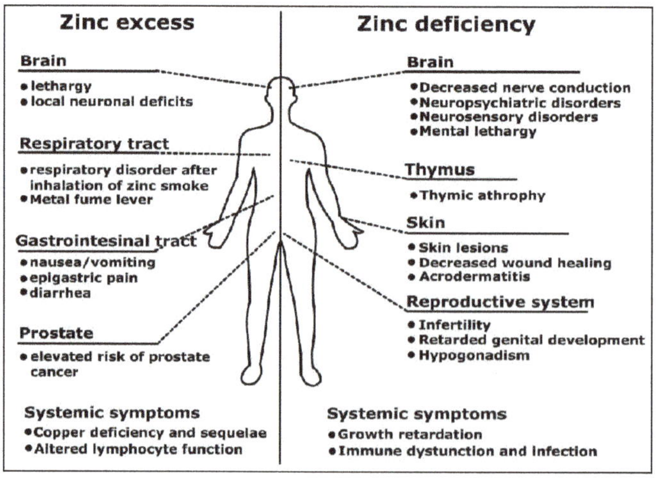

Effects of deficiency and excess of zinc.

The element zinc occurs in over 80 proteins and enzymes, mainly peptidases and anhydrases. The metallothioneins are zinc storage proteins which are useful in combating Cd and Pb poisoning. The common problem with zinc is having a deficiency. A daily consumption of < 5 mg will produce deficiency symptoms, such as dwarfism, dermatistis, loss of taste, immature gonads and delayed wound healing.

Role of Metal Ions in Biological Systems

Metal ion plays an important roles in biological systems either as metallobiomolecules or sometimes it exists as such.

Non-protein Metallobiomolecules – Photoredox

Radiant solar energy enters biosphere using magnesium porphyrin complex (chlorophyll). The process is also referred as photosynthesis and almost all life depend upon chlorophyll and photosynthesis. The oxygen in the atmosphere is a by-product of photosynthesis, and food – stuffs are either parts of plants, or animal which feed on plant.

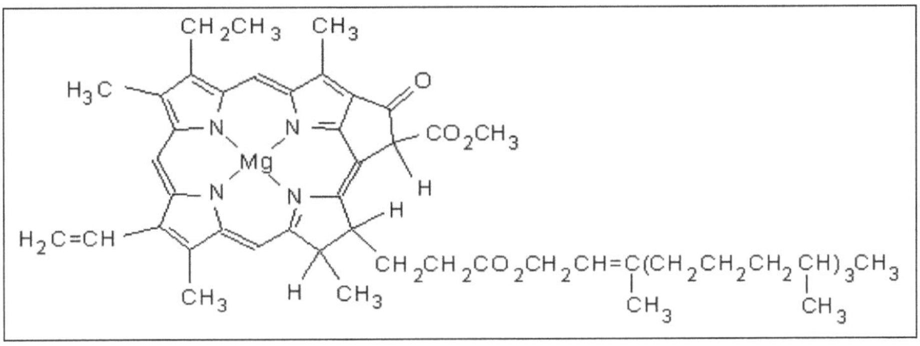

Structure of chlorophyll.

Magnesium also plays an important role in muscle contraction and nerve irritability.

Non-Protein Metal Transport and Structural Metallobiomolecules

The Bacterial Iron Transport

The agents responsible for iron transport into and within bacteria are known as siderophores. They are special chelating agent and are produced by bacteria and ejected into their environment together and transfer iron through the cell wall into the cell. The main problem which is encountered in iron transport is that it is not readily available to aerobic organisms in aqueous environment because of very low solubility product of ferric hydroxide at pH 7. Simple diffusion into cell could never fulfill the requirement of iron inside the cell. Siderophores form chelates with iron and transport the iron into the cell. In some cases it has been seen that siderophore releases the iron at the cell wall and iron passes through alone whereas in other cases entire complex coordination compound enters the cell and releases iron when required.

Skeletal

- Calcium is present in the body in large amounts than any other mineral elements. About 99% of the body calcium is in the skeleton, where it is present as deposits of calcium phosphate in a soft, fibrous matrix. The major inorganic constituent of bone is comprised of a crystalline form of calcium phosphate.
- Silicon appears to be an integral component of acid mucopolysaccharides and may have structural role in connective tissue, cartilage, skin and bone. Silicon is an essential element for growth and skeletal development in chick and rat. It is used in the form of SiO_2 for structural purposes in diatoms, some protozoa, some sponges and one family of plants. Synthetic compounds such as silicones are also becoming very important and are finding many applications in medical science as these can be used as artificial skin etc.
- Phosphorous is found in every cell of the body, but most of it (about 80% of the total) is combined in the bones and teeth. About 10% of it is in combination with proteins, lipids, and carbohydrates, and in other compounds in blood and muscle. The remaining 10% is widely distributed in various chemical compounds. The great importance of the phosphate ester in energy transfer is also known.

Proteins – Transport and Storage Protein Metallobiomolecules

Electron Carriers

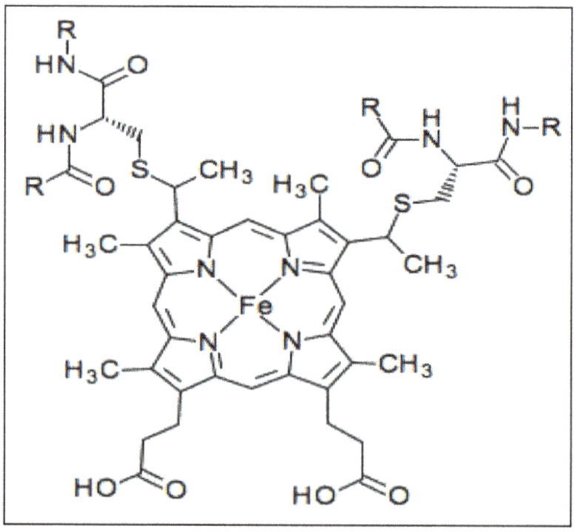

Structure of Cytochrome.

1. Cytochrome: Cytochromes are heme proteins that act as electron carrier and are involved in photosynthesis. It consists of a porphyrin ring chelated to iron atom. Cytochromes are found in all aerobic forms of life. They are not only present in chloroplasts

for photosynthesis but also in mitochondria and take part in reverse process of respiration. It is important to note that iron functions well in cytochromes because Fe^{2+}/Fe^{3+} redox couple emf has a value in the appropriate range for life processes.

2. Iron: Sulphur Proteins: Several non-heme iron sulphur proteins are involved in electron transfer. They are made up of peptide chains having cysteine (amino acid) moiety where cysteine sulphur is coordinated to iron or cluster of iron sulphur atoms. The term rubredoxin is used for those having one Fe atom while those containing Fe- S clusters are called ferredoxin. Ferredoxin is the first electron acceptor in photosynthesis.

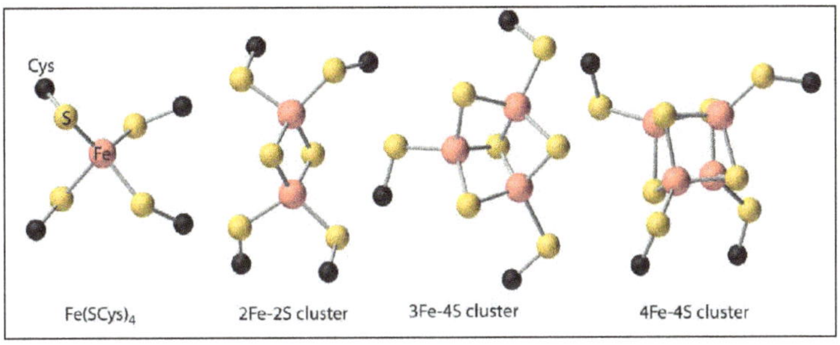

Different types of Iron sulphur proteins.

3. Blue copper: It also acts as an electron carrier in photosynthesis and mainly occurs in algae, green leaves and other plants. The different forms of blue copper are Azurin, Plastocyanin, Stellacyanin and Hemocyanin.

Metal Storage and Structural Metallobiomolecules

1. Ferritin: The regulation of iron absorption is largely carried out by the intestinal epithelial cells. The excess of iron can be stored with cell in non – toxic form by binding to protein called apoferritin in the mucosal cells. Iron(II), once within the mucosal cells, is oxidized to iron(III) and then after combining with apoferritin form iron – protein called ferritin. Therefore, we can say iron can be stored as ferritin and can be released to usable form when required.

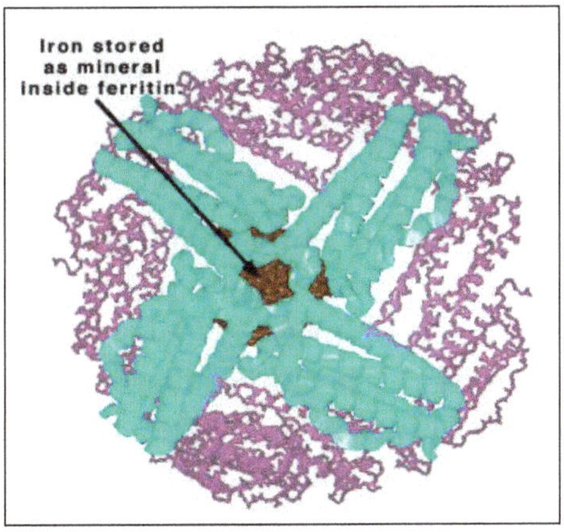

Iron in ferritin.

2. Transferrin: Nearly all the iron released from the mucosal cell enters the portal blood, mostly in iron(II) state. In the plasma (pH is 7.4), iron(II) is oxidized to iron(III) and then picked up by iron binding protein, transferring. This protein is a glycoprotein of MW 77000 containing two similar but not identical sites that bind two atoms of iron(III) per molecule of protein to form a red ferric-protein complex.

3. Ceruloplasmin: It is blue copper binding plasma protein which exerts a catalytic activity (serum ferroxidase) in plasma to convert iron(II) into iron(III) and therefore this metalloenzyme helps in iron incorporation into transferrin.

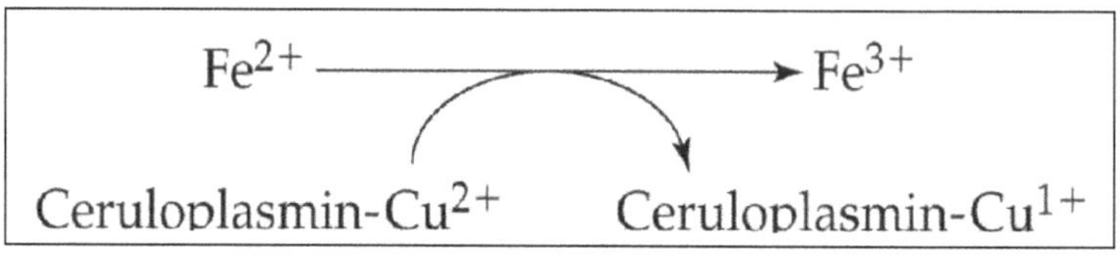

Ceruloplasmin converting Fe(II) to Fe(III).

Oxygen Binding Metallobiomolecules

1. Hemoglobin and Myoglobin: The function of oxygen transport and storage in higher animals is provided by iron – porphyrin complexes, hemoglobin and myoglobin. The former transport the oxygen from lungs to site of its use inside the muscle cell and there the oxygen is transferred to myoglobin for use in respiration. The iron in hemoglobin and myoglobin is in +2 oxidation state. Oxygen molecules coordinates to Fe(II) from one end of heme and globin protects the Fe(II) to be oxidized to Fe(III) by creating a hydrophobic environment so that water molecule could not reach the site. Assuming there is no globin present, the oxidation reaction would have occurred and the product Hematin in which iron is in +3 oxidation state must form. Hematin cannot bind O_2, thus becomes useless for oxygen transport.

2. Hemerythrin: It is a non-heme, oxygen binding iron protein used in great variety of marine worms and phyla of marine invertebrates. It also contains iron(II) which binds oxygen reversibly and when oxidized to iron(III) (methemerythrin) it shows no tendency to bind oxygen. Two forms are known octameric form and monomeric form. Each oxygen binding site contains two iron atoms and reaction takes place via redox reaction to form iron(III) and peroxide (O_2^{2-}) in both monomeric of octameric forms.

3. Hemocyanin: Hemocyanin contain copper and bind one molecule of oxygen for every pair of copper(I) ions. When oxygen is bound to hemocyanin its colour is blue otherwise unbound hemocyanin is colourless. They are found only in molluscs and arthropods (crustacea and arachnida). They occur dissolved in blood or hemolymph and usually comprise of nearly 95% of total protein present. Each subunit contains two atoms of copper and polypeptide chains surrounding them (M.W. 50000-74000). These subunits further form aggregates with a M.W. up to 9×10^6.

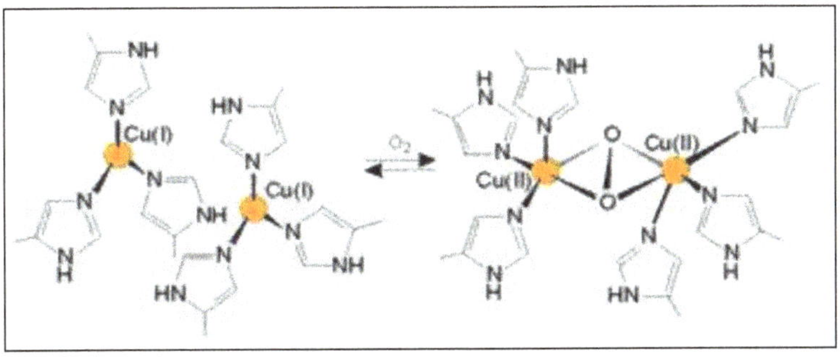

Structure of Hemerythrin.

4. Vacuoles in the blood cells (vanadocytes) of the ascidians contain a vanadium protein complex dissolved in acidic medium. It has been established that carefully prepared vandocytes uptake oxygen reversibly.

Protein – Enzyme Metallobiomolecules

There are many different ways whereby enzymes accelerate the rate of reactions. Metal ions perform catalytic and structural roles in proteins as discussed earlier. Enzymes form metalloenzymes, ternary enzyme metal substrate complexes, in different ways thereby catalyzing different reactions.

Metal ions may play important roles in each of four different mechanisms by which enzymes are known to accelerate the rates of chemical reactions (a) general acid base catalysis, (b) covalent catalysis (c) approximation of reactants and (d) induction of strain in the enzyme or substrate.

Metal ions have ability to accept electrons and therefore these can act as electrophiles. From filled orbitals may also act as electron donor though to a limited extent, therefore sometimes may also act as a nucleophile. Metal ions are supposed to be excellent Lewis acids. Because of all these properties all metal ions have tendency either to form a σ bond or multiple bonds. In addition metals can serve as a three – dimensional template for the orientation and binding of basic groups present on the enzyme or substrate.

Metal ions accepting or donating electrons can activate either electrophiles or nucleophiles and therefore can act as acid base catalyst. Examples include: (i) Kinases, lyases, pyruvate decarboxylase where metal activates an electrophile; (ii) Carbonic anhydrase where metal activates the nucleophile; (iii) Pyruvate carboxylase, carboxypeptidase, alcohol dehydrogenase where metal withdraws π-electrons; (iv) Non-heme iron proteins where metal donates π-electrons.

Sometimes the coordination sphere of a metal may bring together enzyme and substrate or form chelate producing distortion in either enzyme or substrate e.g. in phosphotransferases, D-xylose isomerase and hemoproteins metal ion produces strain effects to accelerate the rate of reaction. Stereochemical control can be achieved by presence of metal ions in enzymes such as pyruvate kinase, pyruvate carboxylase, adenylate kinase when metal ions gather and orient ligands. This is achieved by ability of the metal coordination sphere to act as a three dimensional template to hold reactive groups in a specific steric orientation.

Hydrolases

These are the enzymes catalyzing hydrolysis of ester, ether, peptide, glycosyl, acid anhydride, carbon – carbon bond, carbon – halogen bond or phosphorous – nitrogen bond.

1. Carboxypeptidase: It is a zinc containing metalloenzyme which catalyzes the hydrolysis of the terminal peptide bond at the carboxyl end of peptide chains.

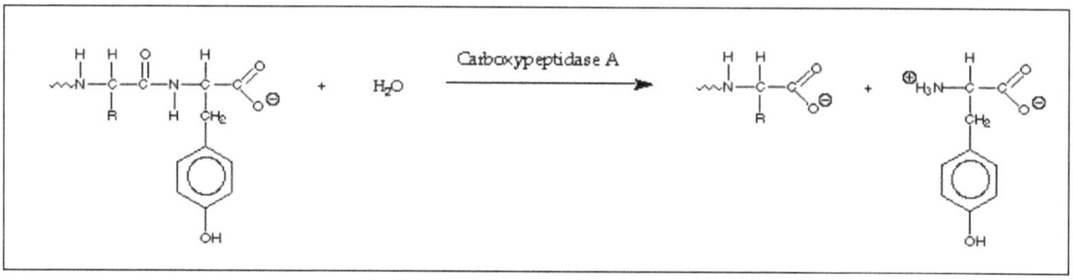

Hydrolysis using carboxypeptidase.

2. Phosphatase: It is a Zn, Cu, Mg containing enzyme and act upon organic phosphates to hydrolyse these into free phosphate.

3. Aminopeptidase: It is a Mg, Mn containing enzyme and hydrolyses polypeptides at the free amino end of the chain to form lower peptides and sometimes free amino acids.

Oxidoreductases

1. Oxidases: There are the enzymes those catalyse the removal of hydrogen from a substrate but use only oxygen as hydrogen acceptor. They invariably contain Cu and form H_2O as a reaction product. Two important examples of this class are (i) cytochrome oxidase (ii) phenolase.

2. Nitrogenase: The evidences available on bacterial nitrogenase point to the activation of molecular nitrogen by metal ions molybdenum and non-heme iron and acid labile sulphide groups are thought to be essential for bacterial nitrogen fixation. The nitrogenase of C. Pasteurianum, A. Vinelandii exhibits hydrogenase activity towards triply bonded substrates. It appears that molybdenum participate in the action of these enzymes in the form of cofactors.

3. Hydrogenases: Some nicotinamide coenzyme dependent dehydrogenases have been found containing zinc. A hydrogenase is an enzyme that catalyses the reversible oxidation of molecular hydrogen (H_2), as shown below:

$$H_2 + A_{ox} \rightarrow 2H{+}+ A_{red}$$
$$2H{+}+ D_{red} \rightarrow H_2 + D_{ox}$$

4. Superoxide dismutase: During aerobic metabolism toxic free radical (O_2^-) is generated and it is needed to remove this free radical from the system. Cytosolic superoxide dismutase helps in converting superoxide radical into water and oxygen and works as an enzyme.

5. Oxygenase: These are the enzymes that catalyse the direct transfer and incorporation of oxygen into a substrate molecule. Oxygenases do not take part in reactions. These provide energy to the cell rather than involving in synthesis and degradation of many different metabolites.

6. Dehydrogenases or Reductases: There are two kinds of dehydrogenases known as aerobic and anaerobic.

Isomerases and Synthetases

These enzymes catalyse interconversion of optical, geometrical or position isomers. Most important example falling under this category is vitamin B-12 coenzyme.

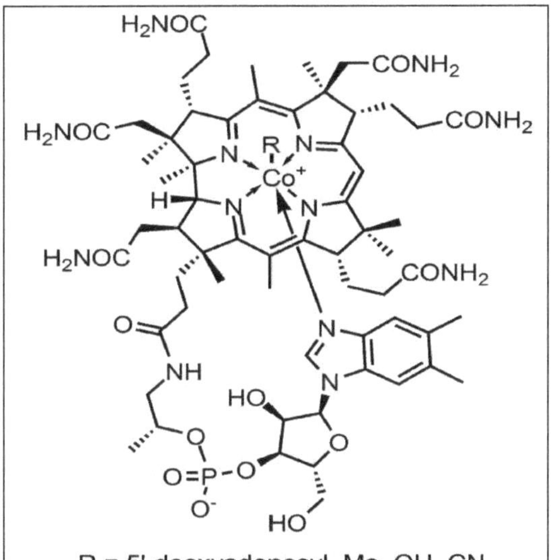

Vitamin B-12 coenzyme structure.

Calcium

1. Participates in the structure of bones and teeth: At least 99% of the total amount of calcium is in the bones and teeth. Calcium in the bones is transformed into a hydroxyapatite - $Ca_{10}(PO_4)_6(OH)_2$ but the bones also contain significant amounts of calcium phosphate, carbonate, citrate, fluoride, magnesium, strontium, trace and minor amounts of other salts. Minerals account about 50% of the total bone mass. The rest is organic matrix in which are proteins, glycoproteins and proteoglycans that bind calcium salts. Every day in the bones can be exchanged up to 700 mg of calcium. The immediate source of calcium for bones is calcium from the body fluids and cells. Although

this amount of calcium is small (<10 g) compared to the amount of bone it is critically important for the regulation surprisingly large number of cellular activities.

2. Metabolic regulation: This is one of the main roles of calcium. Protein kinases, which modulate the activity of key enzymes as a response to hormone binding to the surface of cells, are calcium activated - either directly, or binding to a protein that binds calcium - calmodulin.

3. Regulation of cellular activity: These include nerve and muscle function, hormone action, blood coagulation, cell motility, and many others. Regulates muscle contraction by regulating the contractility of actin and myosin. Since it participates in such a large number of cellular regulation it is also called the second messenger.

4. Intermediates in the reaction of cells to various stimuli: The way of this action is analogous to the regulatory actions of cyclic nucleotides. Effects of calcium are mediated by an intracellular receptor protein - calmodulin, which binds calcium ions when their concentration in the reaction to the stimulus increases. Calmodulin has been found in every studied cell that contains the nucleus. When Ca^{2+} binds to calmodulin it activity many enzymes, and among them are those who participate in the metabolism of cyclic nucleotides, protein phosphorylation, secretory function, muscle contraction, the formation of microtubules, glycogen metabolism and calcium flux. It was found that potent inhibitors of calmodulin activity are phenothiazine drugs that relax the smooth muscles of several peptides that are found in poisonous insects.

5. It is the part of many metalloenzyme: For example, α-amylase and phospholipase contain calcium as an essential part of the catalytic. Osteocalcin is a protein from the bones which is important for normal bone mineral crystallization. Calbindin D is essential for intestinal absorption of calcium, the translation of calcium into the cells and the absorption of calcium from the glomerular filtrate in the kidney. Some of the blood proteins must bind calcium for their activity. Many anticoagulants bind calcium chelate structures (such as EDTA and citrate).

In the calcium-binding proteins calcium effect on:

1. Secretion of hormones and neurotransmitters: Annexin protein must bind to calcium tu bind to a phospholipid membrane. On this case it initiates cellular secretory vesicles to fuse with the surface of membrane and then exocytosis.

2. Cell adhesion: Kaderine are calcium-dependent proteins that regulate cell adhesion and normal contact inhibition of cell replication. Defect in the function of kaderine is linked to the development of malignancy.

3. Cytoskeleton proteins: The importance of Ca^{2+} in these activities is reflected in the precision which regulates the concentration of Ca^{2+} concentrations. Normal plasma contains 9-11 mg of calcium per 100 mL. Daily variations are rarely greater than ±3%.

These narrow limits reflect the complex regulatory action of vitamin D, parathyroid hormone, calcitonin and other hormones.

Magnesium

Magnesium is a Group 2 (alkaline earth) element within the periodic table and has a relative atomic mass of 24.305 Da, a specific gravity at 20 °C of 1.738, a melting point of 648.8 °C and a boiling point of 1090 °C. In the dissolved state, magnesium binds hydration water tighter than calcium, potassium and sodium. Thus, the hydrated magnesium cation is hard to dehydrate. Its radius is ~400 times larger than its dehydrated radius. This difference between the hydrated and the dehydrated state is much more prominent than in sodium (~25-fold), calcium (~25-fold) or potassium (4-fold).

Consequently, the ionic radius of dehydrated magnesium is small but biologically relevant. This simple fact explains a lot of magnesium's peculiarities, including its often antagonistic behavior to calcium, despite similar chemical reactivity and charge. For instance, it is almost impossible for magnesium to pass through narrow channels in biological membranes that can be readily traversed by calcium because magnesium, unlike calcium, cannot be easily stripped of its hydration shell. Steric constraints for magnesium transporters are also far greater than for any other cation transport system: proteins transporting magnesium are required to recognize the large hydrated cation, strip off its hydration shell and deliver the bare (i.e. dehydrated) ion to the trans membrane transport pathway through the membrane. There are obvious chemical similarities between calcium and magnesium but in cell biology, major differences often prevail.

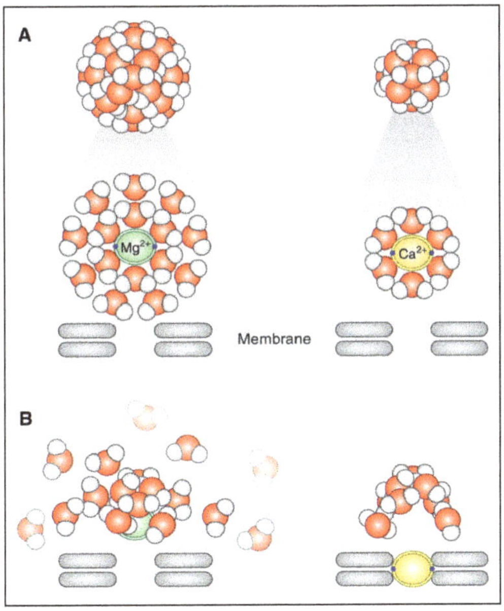

In above figure, (A and B) Magnesium (top left) is surrounded by two hydration shells, whereas calcium (top right) has just one layer. If elements need to fit into a structure (transporter or membrane 'pore'), calcium (below right) simply sheds its hydration shell and its dehydrated ion will fit. Magnesium (below left), on the other hand, first has to get rid of two layers, which is highly energy consuming (simplified model).

Table: Comparison of magnesium and calcium differences and similarities.

	Magnesium	Calcium
Chemical aspects		
Name (symbol)	Magnesium (Mg)	Calcium (Ca)
Element category	Alkaline earth metal.	Alkaline earth metal.
Abundance	Eighth most abundant element in the crust of the Earth.	Fifth most abundant element in the crust of the Earth.
Atomic number	12	20
Valence	2	2
Crystal structure	Hexagonal	Face-centered cubic
Atomic radius	0.65 Å	0.94 Å
Atomic weight	24.305 g/mol	40.08 g/mol
Specific gravity	1.738 (20 °C)	1.55 (20 °C)
Number of hydration shells	Two layers	One layer
Radius after hydration	~400 × larger than its dehydrated form	~25 × larger than its dehydrated form
Isotopes	Magnesium naturally exists in three stable isotopes: Mg (most abundant isotope) Mg Mg Mg radioactive, β-decay	Calcium has five stable isotopes: Ca (most abundant isotope) Ca Ca Ca Ca
Physiological aspects		
Availability in the human body	Normal serum concentration range: 0.65–1.05 mmol/L, divided into three fractions: Free, ionized (ultrafilterable fraction): 55–70%. Protein-bound (non-ultrafilterable): 20–30%. Complexed (citrate, bicarbonate, phosphate): 5–15%.	Normal serum concentration range: 2.2–2.6 mmol/L, divided into three fractions: Free, ionized (ultrafilterable fraction): 47.5–50% Protein-bound (non-ultrafilterable): 42–46% Complexed (citrate, bicarbonate, phosphate): 6.0–6.5%
Total body content in adults	~24 g	~1000 g
Function with respect to cell death	Anti-apoptotic	Pro-apoptotic
Information attained by serum level	Serum level does not represent total body content.	Serum level does not represent total body content.

Physiological Role of Magnesium in the Body

The body of most animals contains ~0.4 g magnesium/kg. The total magnesium content of the human body is reported to be ~20 mmol/kg of fat-free tissue. In other words, total magnesium in the average 70 kg adult with 20% (w/w) fat is ~1000 to 1120 mmol or ~24 g. These values should be interpreted with caution, however, as analytical methods differ considerably throughout the years. In comparison, the body content of calcium is ~1000 g (i.e. 42 times greater than the body content of magnesium).

Distribution in the Human Body

About 99% of total body magnesium is located in bone, muscles and non-muscular soft tissue. Approximately 50–60% of magnesium resides as surface substituents of the hydroxyapatite mineral component of bone. Most of the remaining magnesium is contained in skeletal muscle and soft tissue. The magnesium content of bone decreases with age, and magnesium stored in this way is not completely bioavailable during magnesium deprivation. Nonetheless, bone provides a large exchangeable pool to buffer acute changes in serum magnesium concentration. Overall, one third of skeletal magnesium is exchangeable, serving as a reservoir for maintaining physiological extracellular magnesium levels.

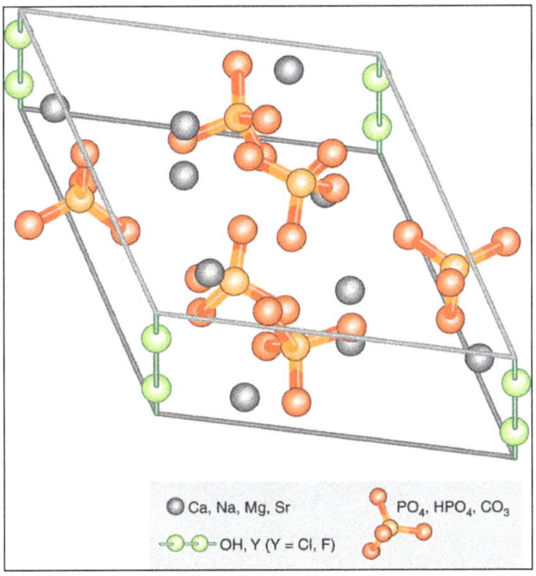

Hydroxyapatite crystal unit: Enamel apatite contains the lowest concentrations of carbonate and magnesium ions, and is rich in fluoride F. Dentin and bone have the highest levels of carbonate and magnesium ions, but have low fluoride content. Fluoride decreases solubility and increases chemical stability, carbonate, chloride and especially magnesium all increase solubility of the otherwise very insoluble mineral. Chemically the mineral comprises a highly substituted carbonated calcium hydroxyapatite (HAP). In the absence of exact compositional analysis the biogenic forms of this mineral are collectively alluded to as "bioapatite". Ca, calcium; Na, sodium; Mg, magnesium; Sr, strontium; OH, hydroxide; Cl, chloride; F, fluoride; PO_4, HPO_4, phosphate; CO_3, carbonate.

Table: Distribution of magnesium in the adult human being, molar mass of magnesium = 24.305 g/mol.

Tissue	Body weight (kg wet weight)	Concentration (mmol/kg wet weight)	Content (mmol)	% of total body magnesium
Serum	3.0	0.85	2.6	0.3
Red blood cells	2.0	2.5	5.0	0.5
Soft tissue	22.7	8.5	193.0	19.3
Muscle	30.0	9.0	270.0	27.0
Bone	12.3	43.2	530.1	52.9
Total	70.0	64.05	1000.7	100.0

Intracellular magnesium concentrations range from 5 to 20 mmol/L; 1–5% is ionized, the remainder is bound to proteins, negatively charged molecules and adenosine triphosphate (ATP).

Extracellular magnesium accounts for ~1% of total body magnesium which is primarily found in serum and red blood cells (RBCs). Serum magnesium can—just like calcium—be categorized into three fractions. It is free/ionized, bound to protein or complexed with anions such as phosphate, bicarbonate and citrate or sulphate. Of the three fractions in plasma, however, ionized magnesium has the greatest biological activity.

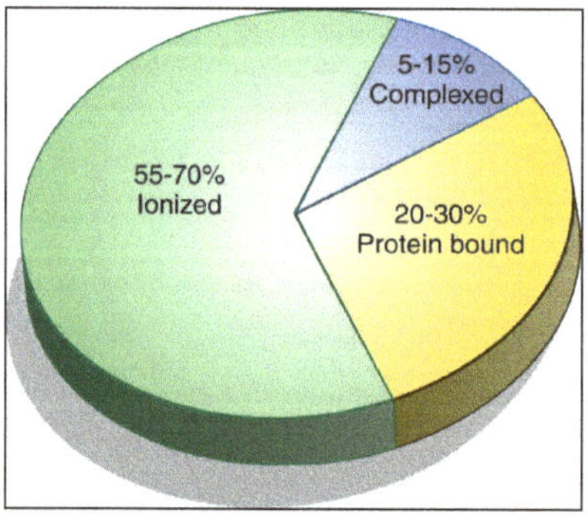

Total serum magnesium is present in three different states. Because of different measurement methods, results published for each state of serum magnesium vary considerably. Therefore, a range for every state is provided.

Magnesium is primarily found within the cell where it acts as a counter ion for the energy-rich ATP and nuclear acids. Magnesium is a cofactor in >300 enzymatic reactions. Magnesium critically stabilizes enzymes, including many ATP-generating reactions. ATP is required universally for glucose utilization, synthesis of fat, proteins, nucleic acids and coenzymes, muscle contraction, methyl group transfer and many other processes, and interference with magnesium metabolism also influences these functions. Thus, one should keep in mind that ATP metabolism, muscle contraction and relaxation, normal neurological function and release of neurotransmitters are all magnesium dependent. It is also important to note that magnesium contributes to the regulation of vascular tone, heart rhythm, platelet-activated thrombosis and bone formation.

Magnesium has numerous functions in the body, for example, serving as a cofactor in enzymatic reactions.

- Enzyme function:
    - Enzyme substrate (ATP-Mg, GTP-Mg),
    - Kinases B,
    - Hexokinase,
    - Creatine kinase,
    - Protein kinase,
    - ATPases or GTPases,
    - Na^+/K^+-ATPase,
    - Ca^{2+}-ATPase,
    - Cyclases,
    - Adenylate cyclase,
    - Guanylate cyclase,
    - Direct enzyme activation,
    - Phosphofructokinase,
    - Creatine kinase,
    - 5-Phosphoribosyl-pyrophosphate synthetase,
    - Adenylate cyclase,
    - Na^+/K^+-ATPase.
- Membrane function:
    - Cell adhesion,
    - Transmembrane electrolyte flux.
- Calcium antagonist:
    - Muscle contraction/relaxation,

    - Neurotransmitter release,
    - Action potential conduction in nodal tissue.
- Structural function:
    - Proteins,
    - Polyribosomes,
    - Nucleic acids,
    - Multiple enzyme complexes,
    - Mitochondria.

In muscle contraction, for example, magnesium stimulates calcium re-uptake by the calcium-activated ATPase of the sarcoplasmic reticulum. Magnesium further modulates insulin signal transduction and cell proliferation and is important for cell adhesion and transmembrane transport including transport of potassium and calcium ions. It also maintains the conformation of nucleic acids and is essential for the structural function of proteins and mitochondria.

It has long been suspected that magnesium may have a role in insulin secretion owing to the altered insulin secretion and sensitivity observed in magnesium-deficient animals. Epidemiological studies have shown a high prevalence of hypomagnesaemia and lower intracellular magnesium concentrations in diabetics. Benefits of magnesium supplementation on the metabolic profile of diabetics have been observed in some, but not all, clinical trials, and so larger prospective studies are needed to determine if dietary magnesium supplementation is associated with beneficial effects in this group.

Recent epidemiological studies have suggested that a relatively young gestational age is associated with magnesium deficiency during pregnancy, which not only induces maternal and fetal nutritional problems but also leads to other consequences that might affect the offspring throughout life.

There is also evidence that magnesium and calcium compete with one another for the same binding sites on plasma protein molecules. It was shown that magnesium antagonizes calcium-dependent release of acetylcholine at motor endplates. Thus, magnesium may be considered a natural 'calcium antagonist'. While calcium is a powerful 'death trigger', magnesium is not: magnesium inhibits calcium-induced cell death. It is anti-apoptotic in mitochondrial permeability transition and antagonizes calcium-overload-triggered apoptosis.

Regulation of Magnesium Influx and Efflux

There is considerable variation in the plasma/tissue exchange of magnesium between various organs of an animal and also between animal species. These observations indicate that various cell types handle magnesium quite differently, which is again different from calcium. Myocardium, kidney parenchyma, fat tissue, skeletal muscle, brain tissue and lymphocytes exchange intracellular and extracellular magnesium at different rates. In mammalian heart, kidney and adipocytes, total intracellular magnesium is able to exchange with plasma magnesium within 3–4 h. In man, equilibrium for magnesium among most tissue compartments is reached very slowly, if at all. About 85% of the whole body magnesium, measured as Mg is either non-exchangeable or exchanges very slowly with a roughly estimated biological half-life of ~1000 h.

Magnesium Consumption

Humans need to consume magnesium regularly to prevent magnesium deficiency, but as the recommended daily allowance for magnesium varies, it is difficult to define accurately what the exact optimal intake should be. Values of ≥300 mg are usually reported with adjusted dosages for age, sex and nutritional status. The Institute of Medicine recommends 310–360 mg and 400–420 mg for adult women and men, respectively. A lower daily minimum intake of 350 mg for men and 280–300 mg magnesium for women (355 mg during pregnancy and lactation).

While drinking water accounts for ~10% of daily magnesium intake, chlorophyll (and thus green vegetables) is the major source of magnesium. Nuts, seeds and unprocessed cereals are also rich in magnesium. Legumes, fruit, meat and fish have an intermediate magnesium concentration. Low magnesium concentrations are found in dairy products. It is noteworthy that processed foods have much lower magnesium content than unrefined grain products and that dietary intake of magnesium in the western world is decreasing owing to the consumption of processed food. With the omnipresence of processed foods, boiling and consumption of de-mineralized soft water, most industrialized countries are deprived of their natural magnesium supply. On the other hand, magnesium supplements are very popular food supplements, especially in the physically active.

Magnesium Absorption and Excretion

Magnesium homeostasis is maintained by the intestine, the bone and the kidneys. Magnesium—just like calcium—is absorbed in the gut and stored in bone mineral, and excess magnesium is excreted by the kidney and the faeces. Magnesium is mainly absorbed in the small intestine, although some is also taken up via the large intestine. Two transport systems for magnesium in the gut are known. The majority of magnesium is absorbed in the small intestine by a passive paracellular mechanism, which is driven by an electrochemical gradient and solvent drag. A minor, yet important, regulatory fraction of magnesium is transported via the transcellular transporter transient receptor potential channel melastatin member (TRPM) 6 and TRPM7—members of the long transient receptor potential channel family—which also play an important role in intestinal calcium absorption. Of the total dietary magnesium consumed, only about 24–76% is absorbed in the gut and the rest is eliminated in the faeces. It is noteworthy that intestinal absorption is not directly proportional to magnesium intake but is dependent mainly on magnesium status. The lower the magnesium level, the more of this element is absorbed in the gut, thus relative magnesium absorption is high when intake is low and vice versa. When intestinal magnesium concentration is low, active transcellular transport prevails, primarily in the distal small intestine and the colon.

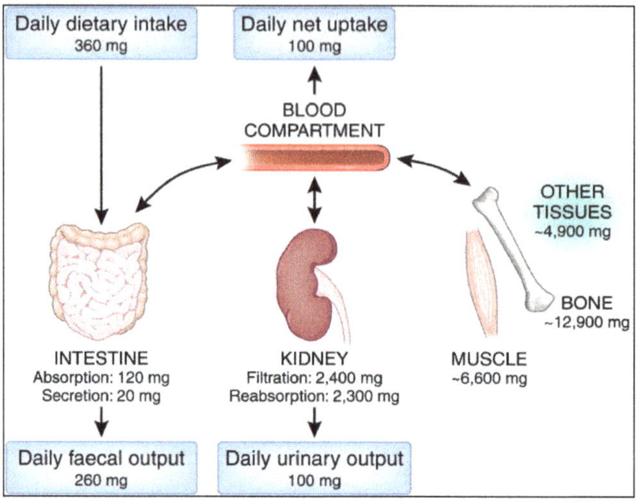

Magnesium balance: Values as indicated based on. The conversion factor from milligrams to millimole is 0.04113.

The kidneys are crucial in magnesium homeostasis as serum magnesium concentration is primarily controlled by its excretion in urine. Magnesium excretion follows a circadian rhythm, with maximal excretion occurring at night. Under physiological conditions, ~2400 mg of magnesium in plasma is filtered by the glomeruli. Of the filtered load, ~95% is immediately reabsorbed and only 3–5% is excreted in the urine, i.e. ~100 mg. It is noteworthy that magnesium transport differs from that of the most other ions since the major re-absorption site is not the proximal tubule, but the thick ascending limb of the loop of Henle. There, 60–70% of magnesium is reabsorbed, and another small percentage (~10%) is absorbed in the distal tubules. The kidneys, however, may lower or increase magnesium excretion and re-absorption within a sizeable range: renal excretion of the filtered load may vary from 0.5 to 70%. On one hand, the kidney is able to conserve magnesium during magnesium deprivation by reducing its excretion; on the other hand, magnesium might also be rapidly excreted in cases of excess intake. While reabsorption mainly depends on magnesium levels in plasma, hormones play only a minor role (e.g. parathyroid hormone, anti-diuretic hormone, glucagon, calcitonin), with oestrogen being an exception to this rule.

- Magnesium is essential for man and has to be consumed regularly and in sufficient amount to prevent deficiency.
- It is a cofactor in more than 300 enzymatic reactions needed for the structural function of proteins, nucleic acids and mitochondria.
- Absorption is complex, depending on the individual's magnesium status, and excretion is controlled primarily by the kidneys.

Sodium

Sodium is one of the body's electrolytes, which are minerals that the body needs in relatively large amounts. Electrolytes carry an electric charge when dissolved in body fluids such as blood. Most of the body's sodium is located in blood and in the fluid around cells. Sodium helps the body keep fluids in a normal balance. Sodium plays a key role in normal nerve and muscle function.

The body obtains sodium through food and drink and loses it primarily in sweat and urine. Healthy kidneys maintain a consistent level of sodium in the body by adjusting the amount excreted in the urine. When sodium consumption and loss are not in balance, the total amount of sodium in the body is affected. The amount (concentration) of sodium in the blood may be:

- Too low (hyponatremia).
- Too high (hypernatremia).

Controlling Blood Volume

The total amount of sodium in the body affects the amount of fluid in blood (blood volume) and around cells. The body continually monitors blood volume and sodium concentration. When either becomes too high, sensors in the heart, blood vessels, and kidneys detect the increases and stimulate the kidneys to increase sodium excretion, thus returning blood volume to normal. When blood volume or sodium concentration becomes too low, the sensors trigger mechanisms to increase blood volume. These mechanisms include the following:

- The kidneys stimulate the adrenal glands to secrete the hormone aldosterone. Aldosterone causes the kidneys to retain sodium and to excrete potassium. When sodium is retained, less urine is produced, eventually causing blood volume to increase.
- The pituitary gland secretes vasopressin (sometimes called antidiuretic hormone). Vasopressin causes the kidneys to conserve water.

Maintaining Fluid and Sodium Balance in Older People

As people age, the body is less able to maintain fluid and sodium balance for several reasons:

- Decreased thirst: As people age, they sense thirst less quickly or less intensely and thus may not drink fluids when needed.
- Changes in the kidneys: Aging kidneys may become less able to reclaim water and electrolytes from the urine (concentrate urine), and, as a result, more water may be excreted in urine.

- Less fluid in the body: In older people, the body contains less fluid. Only 45% of body weight is fluid in older people, compared with 60% in younger people. This change means that a slight loss of fluid and sodium, as can result from a fever or from not eating and drinking enough (sometimes for only a day or two), can have more serious consequences in older people.
- Inability to obtain water: Some older people have physical problems that prevent them from getting something to drink when they are thirsty. Others may have dementia, which may prevent them from realizing they are thirsty or from saying so. These people may have to depend on other people to provide them with water.
- Drugs: Many older people take drugs for high blood pressure, diabetes mellitus, or heart disorders that can make the body excrete excess fluid or magnify the ill effects of fluid loss.

The above situations can result in losing fluid or not consuming enough fluid and thus can cause a high sodium level in blood (hypernatremia) and/or dehydration. Because these situations are more common among older people, hypernatremia is also more common among them. Hypernatremia is poorly tolerated by older people and can result in confusion, coma, and death if severe.

Excess fluid and sodium also occur more commonly in older people because disorders that usually result in excess fluid (fluid overload)—heart failure, liver disorders, and kidney disease—are also more common in older people. A low sodium level in blood (hyponatremia) is more common among older people. Hyponatremia usually results when the body retains too much fluid, as occurs in heart failure or liver disease. Hyponatremia also occurs in older people who take certain types of diuretics (thiazide diuretics such as hydrochlorothiazide), particularly if the kidneys are not functioning normally. Diuretics are sometimes called water pills. Using liquid nutritional supplements or receiving intravenous fluids that are low in sodium while in the hospital also may cause hyponatremia in older people.

Cobalt

Cobalt is very widespread in the natural environment and can be formed as an effect of anthropogenic activity. It is present in small quantities in compounds with sulphur and arsenic. This element is used in numerous industrial applications like welding, diamond tooling, grinding, chemical catalyses, and nuclear power plants. In nuclear plants, it can occur in two radioactive forms of cobalt, ^{58}Co and ^{60}Co, which are formed from corrosion-erosion of alloys containing cobalt and other metals. These two forms produce high-energy γ photons, which can be used in radiotherapy to fight tumours.

Cobalt is used in the production of alloys in plants and in ceramics, as well as chemical catalyses. Moreover, compounds of cobalt are used in industrial production of paints, e.g. Thenard's blue, and lacquers in glass and ceramics.

Organic and Inorganic Cobalt

This essential element occurs in inorganic and organic forms. The first form is essential and necessary for the human body but its excess or deficiency will influence it unfavourably. The organic form of cobalt is present in green parts of plants, fish, cereals, and water. In the human body this element is present in amounts from 1 to 2 mg: we can find it in the heart, liver, kidney, and spleen, and considerably smaller quantities in the pancreas, brain, and serum.

Cobalt is a necessary component of vitamin B^{12} (hydroxocobalamin) and a fundamental coenzyme of cell mitosis. Moreover, cobalt is very important for forming amino acids and some proteins to create myelin sheath in nerve cells. Cobalt also plays a role in creating neurotransmitters, which are indispensable for correct functioning of the organism. The salts of cobalt stimulate the synthesis of erythropoietin, which is the most important function in the activation of different stages of erythropoiesis, which, in turn, is connected with the formation of erythrocytes in bone marrow.

Deficiency of cobalt is strongly related to disturbances in vitamin B^{12} synthesis, so it might cause anaemia and hypofunction of thyroid and increase the risk of developmental abnormalities and failure in infants. However, excess of this metal might increase the action of thyroid and bone marrow, which might, in turn, lead to overproduction of erythrocytes, fibrosis in lungs, and asthma.

For a long time the salts of cobalt have been applied in medicine in the treatment of anaemia, as well as in sport as an attractive alternative to traditional blood doping. This metal enhances synthesis of erythropoietin, which influences enlargement of erythrocyte quantity in blood, improving aerobic performance, but the mechanism of this process is not fully understood. However, it is known that hypoxia is a physiological impulse in erythropoietin production and decreases the oxygen content in circulating blood by bark of adrenal glands and liver.

Inorganic forms of cobalt present in ion form are toxic for the human body. When longer exposed, this element is stored in the body and causes a number of changes in cells. Cobalt chloride and cobalt sulphide are its inorganic forms, which are widely applied in the brewing industry – they are added to beer to prevent the formation of foam. In people consuming large quantities of this drink, cardiomyopathy, a disorder of the diastole of the left ventricle, has been diagnosed. Significant accumulation of cobalt in the heart muscle in these people is formed postmortem. Some derivatives of cobalt, like ^{59}Co, are considered to be carcinogenic and are qualified as such by the International Agency for Cancer Research.

Cobalt in the Human Body

Cobalt daily intake in adults is 3 µg where the content of cobalt is 0.012 µg. Cobalt gets into the body in a few ways. Firstly, with food; secondly by the respiratory system;

thirdly, by skin; and finally, as a component of biomaterials. Biomaterials have already been in use for 2000 years, but their widespread use followed the introduction of polymethyl methacrylate (PMMA) in 1937.

Intake of Cobalt with Food

Cobalt gets into the body with food as shown in figure below, and then it is absorbed in the small intestine. Then the metal gets into the bloodstream where it is bound to proteins and transported with blood to tissues and cells.

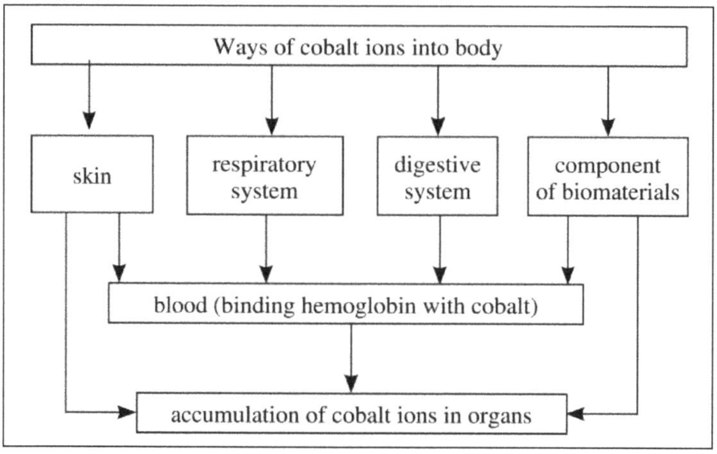

Ways of cobalt ions into body.

The Respiratory System

Substantial intake of cobalt can occur through the respiratory system, especially in people who are employed in heavy metal industry producing and using cobalt metal and cobalt compounds. The lungs absorb a considerable amount of cobalt oxide from dust, welding fumes, metallic cobalt, and other metals. Particles of cobalt and other metals get through to the upper and lower respiratory system and they are displaced depending on their size. The large particles are swallowed and get to the gullet but the smaller ones get through the endothelial cells of the lungs. This situation causes many changes and a risk of lung cancer.

Most epidemiological studies were carried out in workers employed for ten or more years in heavy metal manufacturing plants. In this case lung cancer occurs. Other studies, not epidemiological, demonstrated a connection between inhalation of cobalt and lung cancer. Moreover, chronic cobalt exposure can cause rhinitis, asthma, respiratory irritation, and inflammation. The next studies on animals demonstrated that cobalt exposure can induce many respiratory lesions and alveolar and bronchiolar neoplasms.

Another research project demonstrated that the toxicity of Co ions during the wet grinding of alloy metals can cause a greater risk of allergic alveolitis than in people employed in dry areas with much higher levels of cobalt ions. These cobalt ions are able to induce the production of inflammatory cytokines such as interleukin-6 (IL-6), interleukin-8 (IL-8), and tumour necrosis factor α (TNF-α), by alveolar macrophages as well.

The Skin

The skin is like an external coat and it is susceptible to environmental pollution, especially in workers who are employed in heavy industry. Skin is the biggest immunological organ, which has a very important function in the human body. Its role is to defend the body against pathogens and harmful environmental factors and, secondly, it asserts communication with the external biotope.

The skin is very sensitive when exposed to chemicals and other pollution that can be found in the environment and workplace. Environmental factors, especially heavy metals such as cobalt, can induce allergic contact dermatitis and may cause irritant reactions on the skin. Allergic contact dermatitis affects from 1 to 10 per cent of the total population, and metals such as cobalt, nickel, and chromium can cause contact allergy, affecting 10-15% of women and a few per cent of men.

These diseases can occur especially in workers exposed to more specific occupational conditions, such as in industries preparing hard metals, diamond polishing, and recovering cobalt powders. Cobalt ions from different metal objects and cobalt powders in repetitive contact with the skin can be released from these objects and then they diffuse through the skin and cause allergy and irritant reaction.

Yet another study has demonstrated results of skin absorption of cobalt in industrial workers and in laboratory experiments with volunteers. These investigations were carried out in factories with high levels of pollution. No relation between cobalt ambient air and its urinary secretion was found. On the other hand, when studying heavy metal industry workers whose skin sustained contact with materials containing cobalt, high levels of this metal in serum and urine were detected.

The next experiments were carried out in the laboratory on four people, who were asked to hold their right hands for 90 minutes in a box with freshly mixed wet powder-containing 5-15% cobalt. The level of cobalt was increased by an order of magnitude after exposure and remained unvaried for 48 to 60 hours. Another study evaluated an in vitro system using human abdominal skin obtained from autopsy and synthetic sweat. This research proved that cobalt passed through skin with sweat.

Cobalt gets to the body through the respiratory system and by gastrointestinal or skin absorption, and then it gets into the blood, where is bound to erythrocytes. Sometimes binding of metal by erythrocytes at the normal membrane potential is small but this process could be accelerated by the depolarisation of membrane or the addition of the divalent ionophore cation. This situation brings disorder to electrochemical equilibrium, which could result in extension of cobalt ions intracellularly bound.

The Digestive System

Water-soluble cobalt gets into the body through the gastrointestinal system, and it is absorbed from the small intestine to the blood. Moreover, large quantities of metallic cobalt and its oxide in dust and fumes get into the body through lungs and then they get to the lymphatic and the vascular system. Initially, the concentration of cobalt ions is high but it gradually decreases, reaching a lower level within 24 hours. It is connected with storing cobalt ions in the liver and kidneys and urinary excretion. In another experiment retention of inorganic cobalt in a normal adult human after intravenous injection was studied. The investigation showed that cobalt was reduced by 40% after 24 hours and by as much as to 70% after a week; however, after one month it decreased to about 20% and after one year to about 10%, which means that this amount of cobalt is retained.

Cobalt is accumulated mainly in liver, kidneys, pancreas, heart, skeleton, and skeletal muscles. Moreover, the aim of the investigation was to analyse the mechanism of binding cobalt to red cells ex vivo. The membrane of these cells was depolarised and the results were as follows: the concentration of cobalt ions in cytosol was about 1%, which means that this metal is bound irreversibly to haemoglobin during life, and it circulates in serum penetrating all tissues and organs.

Implants

In America about 500,000 people and in Europe 300,000 people receive joint implants every year. This problem does not concern only old people but also younger ones. Cobalt and its alloys are fundamental components in orthopaedic implants and have been used for about 40 years. Cobalt and other metals, like chromium, molybdenum, and nickel, create alloys that are used as a material to construct total hip arthroplasty.

Prostheses can be made only from metals, and they are called MoM (metal on metal), or they can contain metals and polyethylene – MoP (metal on polyethylene), or only ceramics. Biomaterials are characterised as substances different from a medicament or combination of synthetic or natural substances, and they aim to replace natural body tissues. They are also used in the construction of artificial organs and in the production of medical equipment.

The Biokinetics Model of Cobalt in the Human Body

The model of systemic biokinetics of cobalt was developed for the International Commission on Radiological Protection (ICRP) because of concern for workers who intake radionuclides occurring in the nuclear industry and as diagnostic tools in nuclear medicine. The primary kinetics model was created by Leggett in 1992 and was bound with alkaline earth elements. The structure of the primary model was based on the framework for biokinetics of inorganic cobalt.

This model helps us to understand the biological behaviour of inorganic cobalt in the human body. It was based on many studies carried out earlier on humans and laboratory animals that were exposed to radioactive and also stable cobalt under controlled conditions. The differences between inorganic and organic cobalt were established. Both forms were accumulated in a high concentration in the liver and retained in the body for a long time. This model describes the circulation of cobalt ions between blood and four systemic tissues: liver, kidneys, skeleton, and other organs. The proposed model of behaviour of inorganic cobalt is based on the earlier data by Smith et al. and Letourneau et al. after injecting $^{60}CoCl_2$ and $^{58}CoCl_2$. Measurements were presented as a fractional transfer per day from places where inorganic cobalt was bound and transferred to receptor compartments. This systemic model provides all recycling data as well as for phases of loss as urinary and faecal excretion; some of these results were based on the data from Smith et al. Thus, this model is mainly based on the data by Smith et al. and is supplemented by the data obtained from laboratory animals.

The biokinetics model of inorganic cobalt was created by R. W. Leggett and was based on the research of initial systemic cobalt circulation and its dislocation in compartments of the body. Among these compartments were: liver, kidneys, skeleton, other tissues, and blood. The aim of this biokinetics model was to establish the level of cobalt ions circulating in blood after intravenous injection of ^{60}Co and ^{58}Co. The distribution of cobalt was examine by autoradiography in the body and was expressed in percentages. The blood was divided into blood bound to cobalt ions, which is transferred to organs, and blood remaining in plasma.

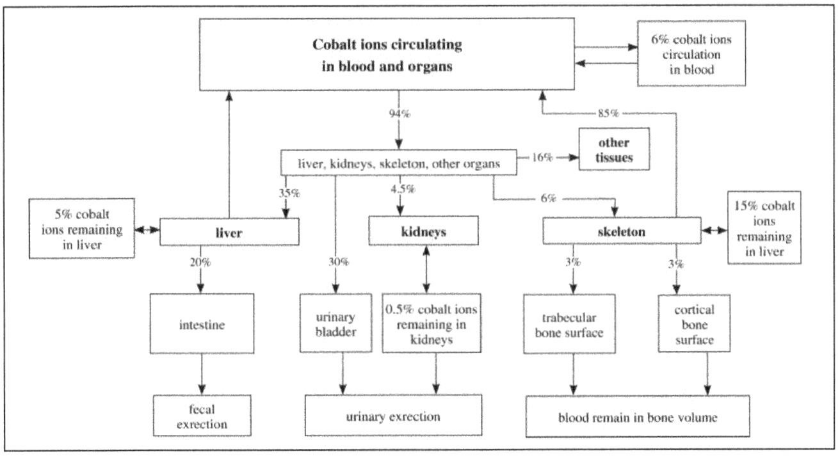

Circulation of cobalt ions between blood and four systemic tissues.

Furthermore, 6% of cobalt ions are bound by blood remaining in plasma, and 94% of cobalt ions are divided among various compartments, including faecal and urinary excretion. Such organs as liver, kidneys, skeleton, and others also represented circulation of cobalt ions in blood. For example, liver received 35% of cobalt ions in blood, and then it was removed from this place in 75% of these ions into plasma. Moreover, the rest of the cobalt ions (20%) contained in bile goes to the small intestine, with 5% remaining and representing the long-term retention of cobalt ions. Kidneys receive 4.5% of cobalt ions in blood and 5% of these ions remain in the kidneys. An additional 30% of cobalt ions go to the urinary bladder from blood that is transferred to organs. Moreover, 6% of ions go to the skeleton, where 3% of them are distributed on the trabecular bone surface and 3% to the cortical bone surface. In addition, 85% of cobalt ions are removed from the skeleton into plasma; 15% of these ions represent long-term retention. Moreover, other organs received 16% of cobalt ions binding in blood.

Application of Cobalt as Biomaterial

The use of biomaterials in medical practice should meet several conditions, e.g. toxicity, biological compatibility, healing of tissues, and mechanical properties. The time of safe use of biomaterials depends on several factors. The most important of them is biocompatibility. It is a property of biological material, which, besides its basic function in the body, does not cause worsening of the patient's condition, or any other complications. Implants with of good compatibility should not be toxic or immunologically active, so they should not induce chronic reactions or an inflammatory condition. In material engineering we can enhance several groups of materials, such as ceramics, polymers, carbon, and metals. All of the groups are different, and they are characterised by different properties.

The metal group contains steel Cr-Ni-Mo, titanium and its alloys, cobalt alloys, niobium, and precious metals. These materials are characterised by various properties such as susceptibility to stretching, brittleness, and mechanical resistance. Due to these properties they are commonly used in the production of various implants, but they are not without defects.

Biomaterials always trigger a reaction in tissues. The biomaterial used in the prosthesis plays a very important role in the success of the implant. Therefore, the use of any biomaterials in the body is determined by the material's biofunctionality and biocompatibility. The latter of these is very important but difficult to determine because of a number of interactions between body substances and biomaterials.

One such interaction is insufficient corrosion resistance in body fluid. The corrosion of metal is the main problem in constructing implants. Several studies have demonstrated that corrosion mechanisms can cause the release of ions of Co, Cr, and Mo into the surrounding tissue and synovial fluid. Moreover, these released metal ions may cause inflammatory and hypersensitivity reactions, and alternations in bone modeling that leads to aseptic loosening and implant failure.

Response of the Human Body to Implantation

The presence of an implant, an alien body in a living organism, detected by the immunological system, triggers certain defensive mechanisms. After the implantation of a prosthesis the first defensive reaction is the creation of biofilm. It means that the surface of the implant is covered with serum albumin and platelets and is defined as an organic or non-organic deposit on the surface of the material and contains cellular elements and bacteria and fungi.

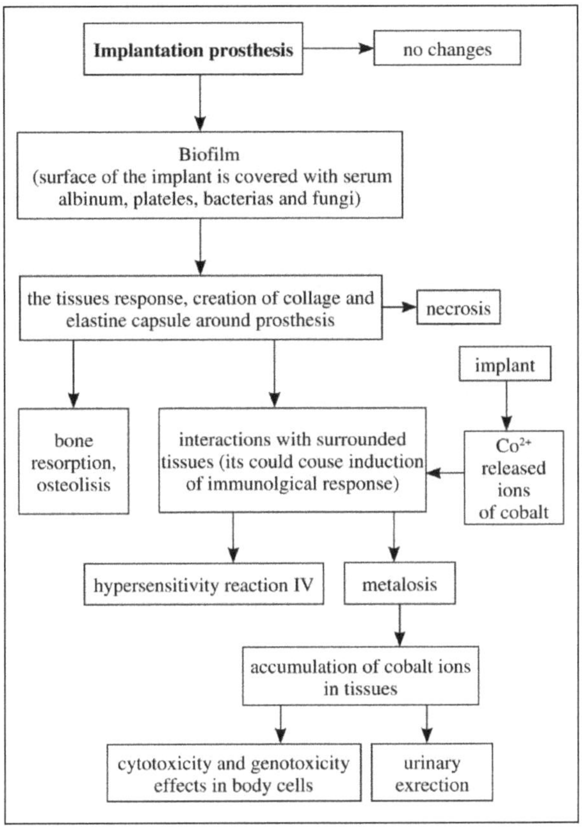

Influence of cobalt ions released from surface of implant and immunological responses of body.

The bacterial forms have the ability to bind extracellular matrix albumins like collagen and fibrinogen, and cause their adhesion (such as: Staphylococcus aureus, Staphylococcus epidermidis, Escherichia coli, Pseudomonas species, Enteroccocus). If the colonisation of bacteria is stopped by the host's cells, such as macrophages and fibroblasts, we cannot observe any changes in the body, but sometimes the defensive mechanism can be weaker than the pace of adhesion, and we can observe the growth of an increasing number of bacteria. The final result of the tissue response is the creation of a collagen and elastin capsule around the prosthesis. This situation could lead to vasa's atrophy and necrosis, bone resorption, and ultimately failure of the implant.

Moreover, this situation could be directly connected with work of implant. The prosthesis contains various alloys of metals such as: cobalt, copper, chromium, molybdenum, and nickel, and they all release ions into the host tissues because of corrosion.

The biological liquids in the body consist of many ions of chlorine, sodium, potassium, lime, phosphates, and organic components, and in addition a higher temperature of the body and changes of pH in tissues surrounding the implant cause the process of corrosion of metals. These ions are released over time, and their level can be highly significant because they are the source of many pathophysiological effects. Studies have demonstrated a sevenfold increase in metal ions in the synovial fluid. Another investigation has shown that Cr and Co ions increase the release of proinflammatory cytokines from macrophages and inhibition of osteoblasts, osteoclasts, and T and B cell proliferation. Moreover, wear debris could be accumulated at the highest doses in the adjacent tissues and bone marrow and could circulate in the bloodstream and penetrate other organs in the body.

The implant is an alien, foreign body, and the organism will activate some mechanisms which aim at its destruction. Macrophages are the type of leukocytes that, after reaching maturity, leave the tissue and get into the bloodstream. These cells are very important components of the immunological system and they synthesize cytokines and other factors that initiate inflammation and bone resorption. All of them cause osteolysis and aseptic loosening and failure of prostheses.

The released ions of metals from metal on the metal surface of periprosthetics could enter the bloodstream and circulate on the body and ultimately accumulate in the heart, liver, kidney, spleen, pancreas, and lymphatic tissue, and only a part of them will be eliminated through urine. This metallic debris may have direct toxicological effects over a long time.

The metallic corrosion products can penetrate the cell plasma membrane, bind cellular proteins or enzymes, and modulate cytokine expression. Osteolysis, and aseptic and non-aseptic loosening are major causes of implant failure.

The reaction of the body to ions of cobalt and other metals is a type IV hypersensitivity reaction. The ions of cobalt and other metals released from the surface of the implant are absorbed by present macrophages, which are involved in many of the processes associated with phagocytose orthopaedic biomaterial particles. If these ions cannot be destroyed after phagocytosis of macrophages then the macrophages resolve and release pro-inflammatory mediators such as IL-1, IL-6, TNF-α, and prostaglandin. The pro-inflammatory cytokines such as IL-1 α andβ stimulate resorption of bone, and then they act synergistically to the tumour necrosis factor TNF-α. Moreover, macrophages release matrix metalloproteinases (MMPs) and chemokines.

The delayed type IV hypersensitivity reaction is a type of immune response where Th helper and cytotoxic Tc cells are also engaged. Th cells are responsible for the damage of infected tissues by macrophages and the activation of cytokines. Another type of cytotoxic Tc provides cell-mediates, which are responsible for cytolysis.

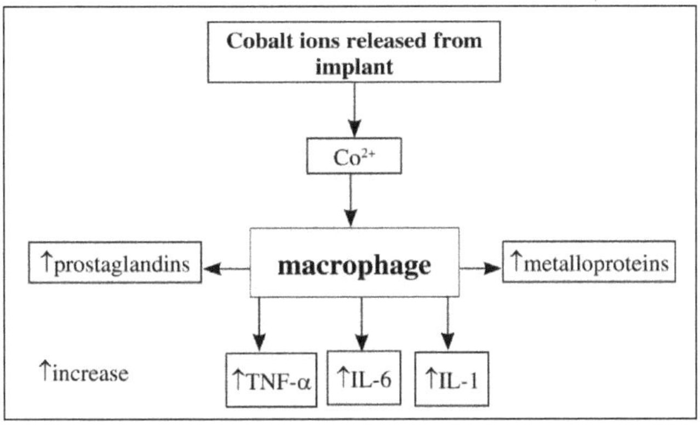

Responses of macrophages on activity of cobalt ions released from implant surface.

Molybdenum

Molybdenum is an essential mineral in the body, just like iron and magnesium. It is present in soil and transferred into your diet when you consume plants, as well as animals that feed on those plants. There is very little data on the specific molybdenum content of certain foods, as it depends on the content of the soil.

Although amounts vary, the richest sources are usually beans, lentils, grains and organ meats, particularly liver and kidney. Poorer sources include other animal products, fruits and many vegetables. Studies have shown that your body doesn't absorb it well from certain foods, particularly soy products. However, this is not considered a problem since other foods are so rich in it. Since your body only needs it in trace amounts and it's abundant in many foods, molybdenum deficiency is rare. For this reason, people don't usually need supplements, unless for some specific medical reasons.

Molybdenum Acts as a Cofactor for Important Enzymes

Molybdenum is vital for many processes in your body. Once you eat it, it is absorbed into your blood from your stomach and gut, then carried to your liver, kidneys and other areas. Some of this mineral is stored in the liver and kidneys, but most of it is converted into a molybdenum cofactor. Any excess molybdenum is then passed in urine. The molybdenum cofactor activates four essential enzymes, which are biological molecules that drive chemical reactions in the body. Below are the four enzymes:

- Sulfite oxidase: Converts sulfite to sulfate, preventing the dangerous buildup of sulfites in the body.
- Aldehyde oxidase: Breaks down aldehydes, which can be toxic to the body. Also, it helps the liver break down alcohol and some drugs, such as those used in cancer therapy.
- Xanthine oxidase: Converts xanthine to uric acid. This reaction helps break

down nucleotides, the building blocks of DNA, when they're no longer needed. They can then be excreted in the urine.

- Mitochondrial amidoxime reducing component (mARC): This enzyme's function isn't fully understood, but it's thought to remove toxic byproducts of metabolism.

Molybdenum's role in breaking down sulfites is especially important. Sulfites are found naturally in foods and also sometimes added as a preservative. If they build up in the body, they can trigger an allergic reaction that can include diarrhea, skin problems or even breathing difficulties.

Chromium

Chromium is among the essential minerals that the body requires to function its best physiologically. Humans require trace amounts of chromium, particularly those who are very active, as chromium and other minerals in their diets ensure that these individuals can sustain increased levels of energy and work.

The predominant form of chromium that is found in the body is trivalent chromium (Cr^{3+}). Cr^{3+}, which can be sourced from many foods, is believed to be involved in normal insulin function. Insulin is key to maintaining the living state and storing carbohydrates, lipids, and proteins within the human body. The element also exists in another form of hexavalent chromium (Cr^{6+}); however, Cr^{6+}, unlike Cr^{3+}, is poisonous and is caused by polluting factories.

Chromium was discovered in the late 1790s; however, it was not until the late 1950s that the biological importance of the element caught the eye of scientists. At that time, researchers found that a component found in Saccharomyces cerevisiae, also known as brewer's yeast, was able to inhibit the aged-based decrease in the ability to preserve regular amounts of blood sugar in rats. Eventually, investigators determined that chromium was the responsible component.

The precise role of a water-soluble compound, which includes chromium and helps insulin maintain proper blood sugar levels, also known as the Glucose Tolerance Factor, is not well understood. However, it is believed that this aids cellular insulin absorption by helping to move insulin across the membranes of a cell. The ability to stimulate insulin absorption has led some to believe that chromium plays an anabolic role in the body.

Chromium Nourishment for the Body

Unlike macronutrients, which are consumed in terms of hundreds of grams each day, micronutrients are consumed daily in terms of micrograms (μg) or milligrams (mg). Chromium helps control whole body metabolism, including how one utilizes energy and how well people function.

In terms of nourishment for the body, nuts, wholegrain cereals, shellfish, eggs, fruits, and vegetables are the primary dietary sources of chromium. In particular, mussels, brazil nuts, oysters, and pears have some of the highest amounts of dietary chromium. Dietary sources of chromium with greater chromium content include:

- Mussels – 128 μg/100g.
- Brazil nuts – 100 μg/100g.
- Oysters – 57 μg/100g.
- Pears – 27 μg/100g.

The body's chromium needs can be readily met by consuming a balanced diet which includes meat, grains, fish, and a variety of produce. Chromium supplements also are available, but there is little evidence to support that any advantages are gained by their use. While there is no recommended dietary allowance for chromium, an adequate and safe intake daily in adults ranges from 50 to 200 μg.

Preventing Disease

Early research in people who had limited tolerance to glucose has found that supplementing their diets with chromium helped glucose utilization or improved their lipid profiles. Some studies have found that supplementation might lower blood sugar levels, as well as their levels of insulin, which is required for those with diabetes.

However, for type 2 diabetes, large-scale randomized controlled trials involving chromium supplementation must be undertaken to determine if chromium can effectively treat the condition. Little evidence exists that could be used to measure the impact that chromium supplementation has on gestational diabetes.

Animal studies have found that blood pressure levels might be lowered with chromium, but the same effect has not been shown to occur in humans. Whether supplementation can lower cholesterol levels in the blood has not been clearly demonstrated.

Overall, ingesting large amounts of chromium does not appear to cause significant negative effects. Despite the fact that the element may be linked to metabolism, investigators must still determine how extensively chromium may impact the body. Among the issues facing them include determining who responds well to supplementation, analyzing a food product's chromium content, and how extensively chromium is taken up by the circulatory system, as well as establishing accurate benchmarks that indicate chromium levels.

Copper

Copper is an essential trace element vital to the health of all living organisms. Although the amount of copper found in the human body (50–120 milligrams) would fit on the head of a pin, this tiny quantity is essential. Research has revealed that copper is vital for the optimal health of the human body, along with other mineral micronutrients such as iron, calcium and zinc.

The human body contains around 2 mg of copper per kilogram of body mass. While copper is found throughout the body, it is concentrated in organs with high metabolic activity, such as the liver, kidneys, heart and brain. This tiny amount is enough to provide copper ions for billions of protein molecules, in particular enzymes, where the copper ions are essential cofactors. Without cofactors, the enzymes could not work.

Copper is an essential trace mineral that cannot be formed by the human body so must be ingested from dietary sources every day. According to the World Health Organization, 1–3 milligrams per day of copper are required to prevent any symptoms of deficit. Various health and nutrition organizations around the world have set dietary reference values, highlighting the importance of copper as part of a balanced diet.

Adequate Copper Intakes (mg/day)		
Age	Female	Male
7–11 months	0.4	0.4
1 – <3 years	0.7	0.7
3 – <10 years	1.0	1.0
10 – <18 years	1.1	1.3
≥18 years	1.3	1.6
Pregnancy	1.5	
Lactation	1.5	

Copper is used Throughout the Body

Virtually every cell in the body utilizes copper and – together with iron and zinc – copper is one of the trio of minerals essential to good health. Copper is vital to the health of the body from foetal development through to old age. Quite simply, without copper our brains, nervous systems and cardiovascular systems could not function normally. Copper is essential for:

- Brain development during foetal and post-natal growth, and maintenance of brain health throughout life, including effective anti-oxidative defense.
- Efficient communication between nerve cells.
- Maintenance of healthy skin and connective tissue.
- Wound healing.
- Structural integrity and function of heart and blood vessels.
- Growth of new blood vessels.
- Proper structure and function of circulating blood cells.
- Formation of the cells of our immune system (white blood cells).
- Maintenance of a healthy and effective immune response.
- Generation and storage of energy in the 'power plants' of our cells, the mitochondria.

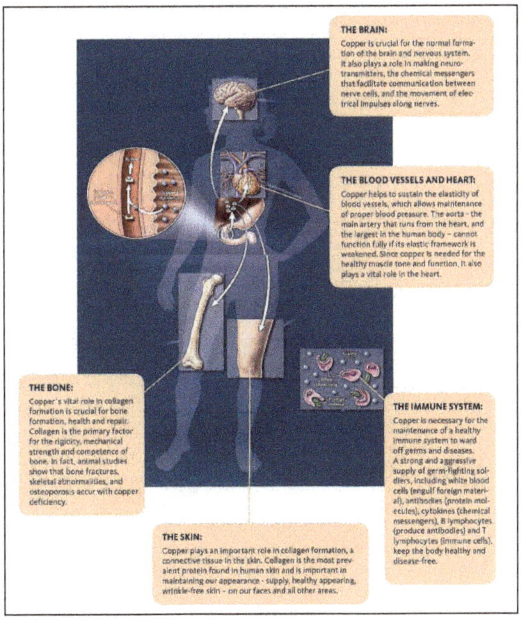

Beyond the Brain

The Liver

After copper is ingested, it is absorbed from the small intestine into the bloodstream.

Here, copper bonds with transport proteins, which carry it to the liver. Once copper is taken up into the liver, it is stored, distributed around the body in transport proteins or excreted into the bile. Copper plays a central role in the conversion of iron to its useable iron (III) (ferric) form and also helps transport iron around the body. Copper deficiency can lead to anaemia and tissue iron overload.

The Brain and Nervous System

Copper is crucial for the normal development of the brain and nervous system. It plays a role in the production and maintenance of myelin, which insulates nerve cells, so ensuring the proper transmission of nerve impulses. Copper is also involved in the synthesis of neurotransmitters, the chemicals that allow communication between nerve cells. Copper deficiency can lead to degeneration of the nervous system.

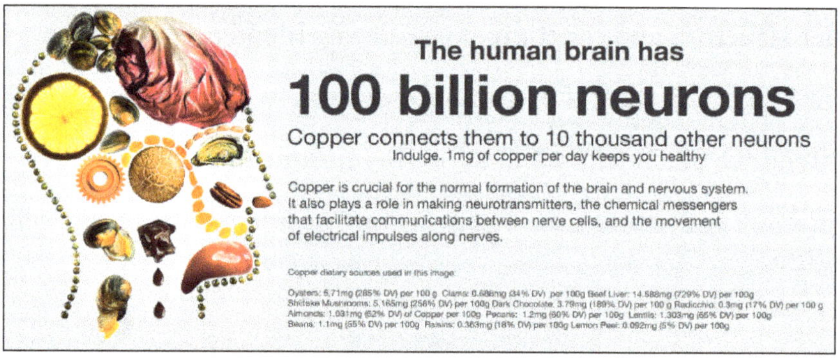

Cells

The release of cellular energy (ATP) inside the mitochondria depends on the crucial involvement of a copper-containing enzyme. A similarly vital function for copper, as a cofactor, is the neutralisation of free radicals that would otherwise oxidise and destroy healthy cells. Copper deficiency can lead to cancer and heart disease.

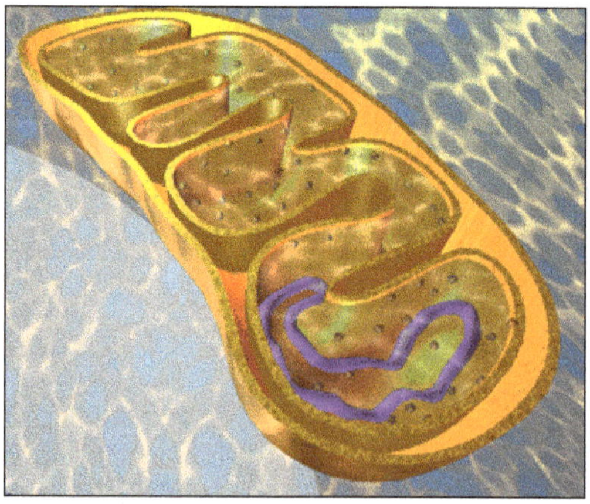

Copper is essential for the reactions in mitochondria that drive the metabolism of cells. This illustration shows a section through a mitochondrion.

The Heart

Copper is essential for the synthesis of collagen. This is found in connective tissue, which is the main supporting and binding tissue of the body. Copper is also needed for healthy muscle tone and function and so plays a vital role in the heart. Copper deficiency can lead to heart failure.

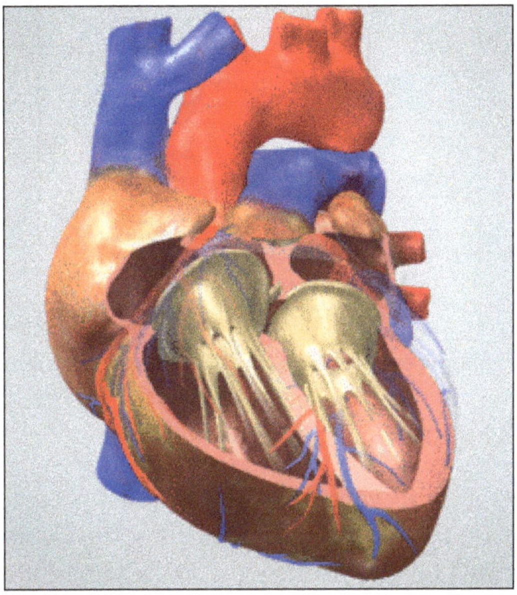

Muscle and connective tissue in the heart contain elastin and collagen. Copper is the cofactor in lysyl oxidase, the enzyme involved in cross linking these molecules to form strong tissue.

Blood Vessels

Copper serves as a cofactor for an enzyme involved in the coagulation of blood. The blood vessels are surrounded and protected by connective tissue, and copper helps to sustain their elasticity, particularly for the aorta and smaller arteries. Copper deficiency can lead to circulatory problems and low blood pressure.

Copper is essential for a wide range of biological functions that play an important role in maintaining healthy skin, namely tone, wound healing, and ultraviolet (UV) protection. Copper is required for the production of collagen and elastin, two components that provide support and elasticity for skin. As a cofactor for the enzyme tyrosinase, copper is involved in the synthesis of the skin pigment melanin, which provides protection from UV radiation. Copper deficiency can lead to skin degeneration and loss of pigmentation.

Bone

Collagen is the main structural material in bone. Cross-linking of collagen molecules affects bone strength. Enzymes with copper cofactors are involved in this process. Copper deficiency can lead to fractures, skeletal abnormalities and osteoporosis.

The Immune System

Copper is necessary for the maintenance of a healthy white blood cell count; many of these white cells are phagocytes which engulf and destroy microorganisms. Copper deficiency can lead to a depressed immune system, reduced white cell count and increased incidence of pneumonia.

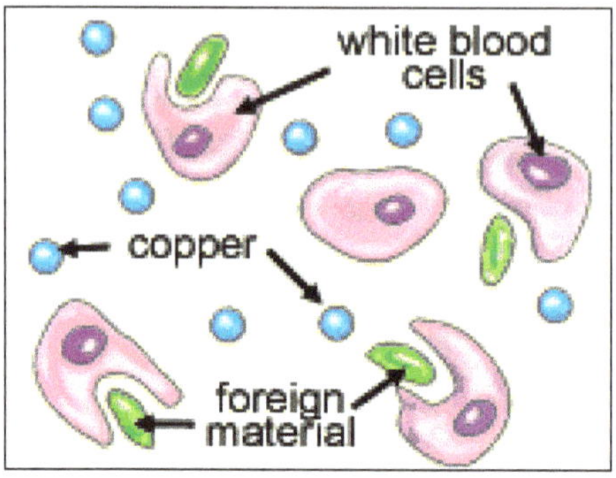

Copper is essential for a healthy immune system.

Dietary Sources of Copper

Copper is available from a wide variety of fresh and processed foods. Good sources include grains, nuts (particularly brazil and cashew nuts), meats (liver and kidney), shellfish, legumes (peas and beans) and seeds. Dark chocolate is also a rich source of copper. Eating a well-balanced diet should allow you to meet your daily requirements of copper.

Some of the copper rich foods. If you don't like lobster, dark chocolate is also a good source.

Copper Deficiency

Copper is an essential micronutrient. Deficiency causes serious health problems, but is very rare. Deficiency disorders are more often caused by genetic faults in the transport proteins than by a diet low in copper. The copper is available, but it cannot get into cells. A healthy balanced diet should provide the copper required each day. There is concern that poor or unbalanced diets may not provide sufficient copper.

Copper Excess

Copper homeostasis balances uptake and excretion of copper to meet the body's needs on a continuous basis. Acute copper poisoning is a rare event, largely restricted to the accidental drinking of solutions of copper nitrate or copper sulfate, which should be kept out of easy access in the home. These, and organic copper salts, are powerful emetics and large doses are normally rejected by vomiting. The capacity for healthy human livers to excrete copper is considerable and chronic copper poisoning is very rare, with the few reports of it restricted to patients with liver disease.

Balancing Trace Elements

The intake of adequate copper is, on its own, not good enough. Minerals interact in the body, working as a team. The advice of research scientists is to regard copper, iron and zinc as an essential trio – too much of one can hamper the body's ability to absorb or process the other two. Deficiency diseases may then result.

Table: Sources of minerals.

Mineral	Function in human body	Deficiency disease	Common food sources
Iron	• Oxygen carrier (haem) in haemoglobin. • Electron carrier (in respiration). • Enzyme cofactor (catalase; peroxidase – breakdown of hydrogen peroxide).	• Anaemia	• Liver • Red meat • Spinach
Zinc	• Enzyme cofactor (carbonic anhydrase – carriage of carbon dioxide; carboxypeptidase – protein digestion).	• Growth retardation • Loss of appetite • Impaired immune function	• Most foods

Deficiencies

Too much zinc can inhibit the absorption of copper. This may result in poor copper status in the body, which over time may lead to bone abnormalities, problems with the heart and circulation and impaired immunity. Iron needs copper and without adequate copper, iron cannot be converted to its most useful iron (III) (ferric) form. Copper is also necessary for the transport of iron to and from the tissues.

Copper is increasingly being recognised as an essential micronutrient. A growing body of research indicates that, as part of a balanced diet, an adequate intake of copper is vital to ensure a wide range of health benefits.

Iron

Iron performs many important functions in the body. It is primarily involved in the transfer of oxygen from the lungs to tissues. However, iron also plays a role in metabolism as a component of some proteins and enzymes.

Iron is toxic to the body in its free state. It is associated with proteins either through ligand binding or by being incorporated into a porphyrin group - a ring-shaped molecule. A complex of the ferrous form of iron and protoporphyrin IX is known as heme. Heme iron is found in proteins connected with oxygen transport, including hemoglobin and myoglobin. Non-heme iron can be found in proteins connected with oxidative phosphorylation and in iron storage proteins like transferrin and ferritin.

Products containing iron: Healthy eating.

Hemoglobin and Myoglobin

About 70 percent of iron in the body is found in hemoglobin and myoglobin. Hemoglobin is the protein in red blood cells responsible for carrying oxygen to the tissues from the lungs. Myoglobin is a protein found in muscles that is used for storage of oxygen. Hemoglobin is the oxygen transport system found in the red blood cells of all vertebrates and some invertebrates. In humans, hemoglobin is made up of four globular protein subunits. The four subunits form a pocket that binds a heme group.

Oxygen binds to the iron atom within the hemoglobin molecule in the lungs to form oxyhemoglobin. This occurs in the capillaries of the lung alveoli. It is released at its destination in the cells. Hemoglobin carries CO_2 back to the lungs to be exhaled as waste, but CO_2 binds to the protein portion of the hemoglobin molecule, not to the bound iron in the heme group.

Like hemoglobin, myoglobin binds iron within a heme group. However, structurally, it is much simpler, consisting of a single polypeptide chain of 154 amino acids. It is found only in cardiac myocytes and oxidative skeletal muscle. Myoglobin is an oxygen storage protein. In marine mammals, it provides an oxygen supply for extended periods when the animal is diving under water. At those times, myoglobin releases oxygen to sustain aerobic metabolism in the muscle. In humans, myoglobin levels have been shown to be increased at high altitudes.

Iron-Dependent Enzymes

A large number of enzymes require iron as a cofactor for their functions. Among the most significant of these are enzymes involved in oxidative phosphorylation, the metabolic pathway that converts nutrients to energy. The cytochrome enzymes bind heme iron, and some protein complexes in the oxidative phosphorylation process have iron-sulfur centers that are crucial to their function.

Ferritin and Transferrin

Dietary iron is stored within a protein complex called ferritin. Ferritin has 24 subunits that form a capsule around the bound iron atoms. Each complex binds 2000 to 45000 iron atoms. Another protein, transferrin, made in the liver, transports iron within the blood to other locations for storage. The main locations of iron storage in the body include the liver, skeletal muscle, and reticuloendothelial cells. If the storage capacity of these cells is exceeded, iron is deposited near the ferritin-iron complexes in the cells. These deposits are called hemosiderin. Iron in hemosiderin is not available to the cell. Hemosiderin deposits may be found in the body following a hemorrhage.

The overall utilization of iron in the body is regulated by ferritin and transferrin mRNAs that contain iron responsive elements (IREs). Iron homeostasis requires ascorbic acid (vitamin C), which stimulates the absorption of dietary iron, and assists with uptake of transferrin-bound iron in the plasma. Ascorbic acid also stimulates synthesis of ferritin, while inhibiting ferritin degradation and flow of iron out of the cell.

Examples of Metal Ions in Biological System

Some more examples involving metal ions and their complexes playing important role are given below:

1. Carbonic anhydrase: It is a metalloenzyme containing one gram ion of Zn (II) per peptide chain, which is essential for enzymatic activity of carbonic anhydrases. The biologically important role of carbonic anhydrase is rapid hydration of metabolic carbon dioxide formed in tissues, dehydration of HCO_3^- in lungs, transfer and accumulation of H_+ or HCO_3^- in organs of secretion.

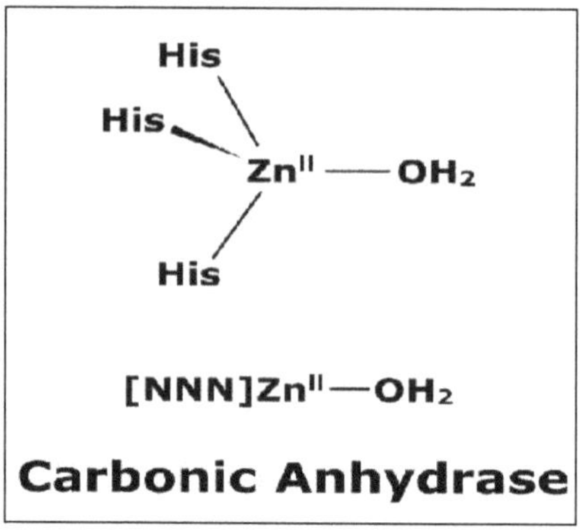

Zinc in Carbonic anhydrase.

2. Catalase and Peroxidase: Both H_2O_2 and superoxide ion are reactive and toxic and may be produced in any biological system by auto oxidation of many organic compounds in vivo. Catalase catalyses decomposition of peroxide or superoxide ion to H_2O and O_2 (disproportionation reaction). Peroxidase catalyses the oxidation of many substrate into peroxide and catalase help in decomposition H_2O_2 into H_2O and O_2. Peroxidase and catalase have many similarities both in structure and mechanism. Both have high spin ferric heme groups surrounded with large protein molecules.

3. Nitrate Reductase: Nitrate is a source of nitrogen for most green plants and fungi. Nitrate is converted into ammonia which is further utilized in synthesis of proteins, nucleic acids and other cell constituents, this process is known as assimilatory nitrate reductions. Nitrate may be used in place of oxygen as electron acceptor in anaerobic production of energy to certain bacteria, this is known as dissimilatory nitrate reduction. A flavoprotein, NADH-nitrate reductase and iron-sulphur protein ferredoxin-nitrate reductase converts nitrate into nitrite by first enzyme and then nitrite is converted into ammonia by second enzyme respectively. Molybdenum is essential for reduction of nitrate to nitrite by assimilatory nitrate reductase from Neurospora.

4. Sulphite Oxidase: The enzymatic catalysed oxidation of sulphite occurs in mammals, plants and bacteria and its role appear to be in oxidative detoxification of sulphur compounds. The prosthetic group of enzyme is heme in most of cases, and it has been shown that two hemes are present per mole out of sulphite oxidase which consists of two subunits. During the oxidation process of $So_3^{2-} \rightarrow SO_4^{2-}$, one heme is reduced rapidly than the second which reduces much slowly. Studies show that it also contains molybdenum and ratio of Mo: heme is 1:1.

5. Xanthine Oxidase: Its composition closely resembles to that of xanthine dehydrogenase which is mostly found in chicken liver. The enzyme contains two molybdenum atom, two FAD molecules and eight iron atom per molecule. Its role is that it helps in oxidizing xanthine into uric acid. It can also catalyse the oxidation of aldehyde to carboxylic acid.

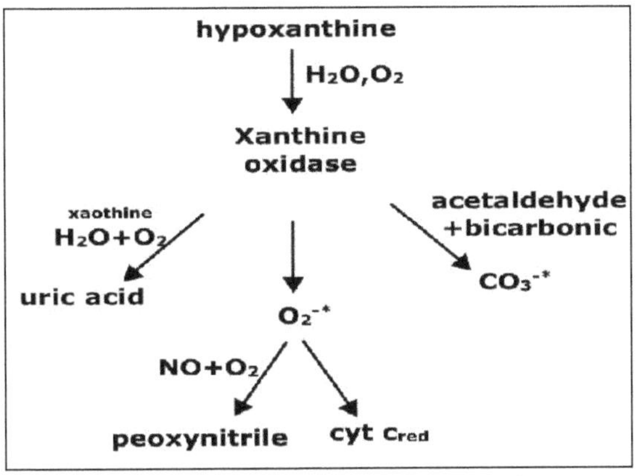

Different reactions by Xanthine Oxidase.

6. Aldehyde Oxidase: This enzyme (rabbit liver) have some similarities with xanthine oxidase. Both have identical contents of Mo, Fe and FAD although aldehyde oxidase also contains two molecules of coenzyme Q. They have ability to utilize molecular oxygen and generating peroxide anion. They have ability to catalyse the hydroxylation of aldehyde and N-heterocyclic substrates.

7. Carp-muscle Albumins: Numerous low molecular weight albumins occurring in the white muscles of fish are known to have biochemical properties such as in mediation of the effect of calcium in muscle contraction.

8. Ca(II) requirement for enzymatic activity is also well known. Staphylococcal nuclease is a single chain polypeptide enzyme of 149 amino acid residue which exhibits hydrolytic activity towards both deoxyribonucleic acid (DNA) and ribonucleic acid (RNA). Enzymatic activity of staphylococcal nuclease appears to be completely dependent upon Ca(II) ion when RNA or DNA is hydrolysed.

It has also been pointed out that high co-ordination number usually associated with Ca(II) and its ability to act as a bridge between anions makes it highly suitable for integration as structural component in biological system. In addition to these catalytic activity, Ca is a major constituent of bones and teeth. It is essential for muscle contraction, normal heart rhythm and nerve irritability.

9. Metallothioneins (cysteine rich proteins) have capabilities to bind metal such as Hg, Ca, Zn and Cu those have certain affinity towards thiolate groups (soft-soft interaction). Seven metal ions can bind to one peptide chain and this makes metallothionein to work as detoxifying agent (especially against Cd and Hg). In addition to this metallothioneins also play important role in metabolic processes.

10. Copper is the third most abundant metallic element in the human body Fe and Zn. In addition to electron transfer reaction in photosynthesis it also helps to (i) oxidize galactose when present in galactose oxidase, (ii) convert dopamine (Adernols) to nose pine phrine when present in Dopamine β-hydroxylase.

11. Urease from jack beans conatin tightly bound nickel which activates the enzyme made up of polypeptide. The main function of urease is to catalyse the hydrolysis of urea to ammonia and carbonic acid. Many different bacteria are known to conatin nickel which act as hydrogenase and catalyse the reaction such as formation of water from molecular hydrogen and oxygen, production of methane, reduction of sulphate ion etc.

12. Glucose tolerance factor contains chromium (III) along with glycine, cysteine, glutamic acid, nicotinic acid and is thought to be an essential trace element.

13. Many of the bacterial enzymes such as formate dehydrogenase contain tungsten.

References

- Toxic-elements: hairanalysisreport.com, Retrieved 10, May 2020
- Bioinorganic-chemistry-notes-lecture-3: arsdcollege.ac.in, Retrieved 23, July 2020
- Overview-of-sodiums-role-in-the-body, electrolyte-balance, hormonal-and-metabolic-disorders: msdmanuals.com, Retrieved 12, April 2020
- How-Chromium-Functions-in-the-Body: news-medical.net, Retrieved 03, February 2020
- Copper-essential-human-health, education-resources, knowledge-base: org.uk, Retrieved 20, June 2020
- Iron-Functions-in-the-Body, health: news-medical.net, Retrieved 08, January 2020

3

Metalloproteins

The proteins which have a metal ion as a cofactor are called metalloproteins. Most of the proteins fall in this category. A few types of metalloproteins are zinc enzymes and iron sulphur proteins. The topics elaborated in this chapter will help in gaining a better perspective about these metalloproteins as well as design of functional metalloproteins.

Metalloproteins are proteins bound by at least one metal ion. Metal ions are usually coordinated by four sites consisting of the protein's nitrogen, sulphur and/or oxygen atoms. In metalloenzymes, one of the coordination sites is labile. The chemistry of metals allows for a broader set of reactions, for instance as in redox reactions.

Metalloproteins are a class of biologically important macromolecules, which have various functions such as oxygen transport, electron transfer, oxidation, and oxygenation. These diverse functions of metalloproteins have been thought to depend on the ligands from amino acid, coordination structures, and protein structures in immediate vicinity of metal ions. In this project, we are studying the relationship between the electronic structures of the metal active sites and reactivity of metalloproteins.

1. One-Electron Oxidation of Electronically-Diverse Manganese(III) and Nickel(II) Salen Complexes: Transition from Localized to Delocalized Mixed-Valence Ligand Radicals).

Ligand radicals from salen complexes are unique mixedvalence compounds, in which a phenoxyl radical is electronically linked to a remote phenolate via a neighboring redox-active metal ion, providing an opportunity to study electron transfer from a phenolate to a phenoxyl radical mediated by a redox-active metal ion as a bridge. We herein synthesize one-electron oxidized products from electronicallydiverse manganese(III) salen complexes, in which the locus of oxidation is shown to be ligand-centered, not metal-centered, affording manganese(III)–phenoxyl radical species. The key point in the present study is an unambiguous assignment of IVCT (intervalence charge transfer) bands by using nonsymmetrical salen complexes, which enables us to obtain otherwise inaccessible insight into the mixed-valence property.

A d^4 high-spin manganese(III) ion forms a Robin-Day Class II mixed-valence system, in which electron transfer is occurring between the localized phenoxyl radical and the phenolate. This is in clear contrast to a d^8 low-spin nickel(II) ion with the same salen ligand, which induces a delocalized radical (RobinDay Class III) over the two phenolate rings, as previously reported by others. The present findings point to a fascinating possibility that electron transfer could be drastically modulated by exchanging the metal ion that bridges the two redox centers.

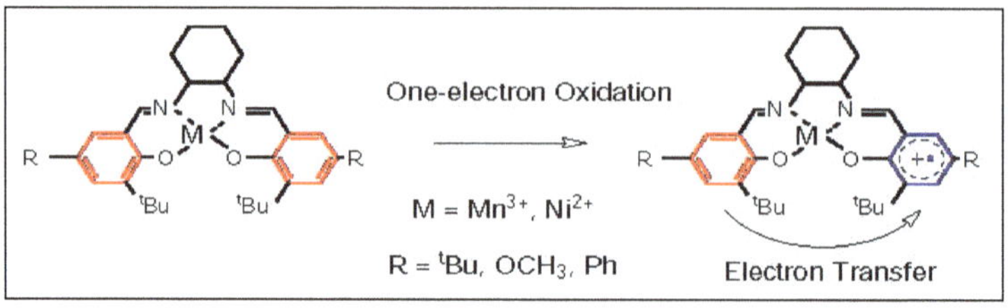

Mixed-Valence Ligand Radicals formed from manganese (III) and nickel(II) salen Complexes.

2. Redox Potentials of Oxoiron(IV) Porphyrin π-Cation Radical Complexes: Participation of Electron Transfer Process in Oxygenation Reactions).

The oxoiron(IV) porphyrin π-cation radical complex (compound I) has been identified as the key reactive intermediate of several heme enzymes and synthetic heme complexes. The redox properties of this reactive species are not yet well understood. Here, we report the results of a systematic study of the electrochemistry of oxoiron(IV) porphyrin π-cation radical complexes with various porphyrin structures and axial ligands in organic solvents at low temperatures. The cyclic voltammogram of (TMP)$Fe^{IV}O$, (TMP = 5,10,15,20-tetramesitylporphyrinate), exhibits two quasi-reversible redox waves at $E_{1/2}$ = 0.88 and 1.18 V vs. SCE in dichloromethane at –60 °C. Absorption spectral measurements for electrochemical oxidation at controlled potential clearly indicated that the first redox wave results from the (TMP) $Fe^{IV}O$/[(TMP+•)$Fe^{IV}O$]+ couple. The redox potential for the (TMP)$Fe^{IV}O$/[(TMP+•)$Fe^{IV}O$]+ couple undergoes a positive shift upon coordination of an anionic axial ligand, but a negative shift upon coordination of a neutral axial ligand (imidazole). The negative shifts of the redox potential for the imidazole complexes are contrary to their high oxygenation activity. On the other hand, the electron-withdrawing effect of the meso-substituent shifts the redox potential in a positive direction. Comparison of the measured redox potentials and reaction rate constants for epoxidation of cyclooctene and demethylation of N,N-dimethylanilines enable us to discuss the details of the electron transfer process from substrates to the oxoiron(IV) porphyrin π-cation radical complex in the oxygenation mechanisms.

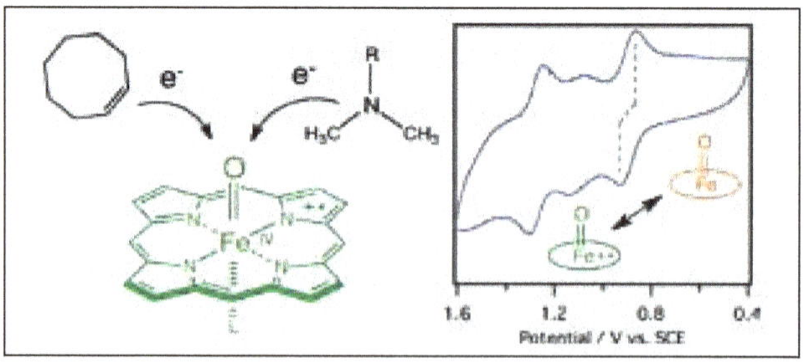

Cyclic Voltammogram of Oxoiron(IV) Porphyrin π-Cation Radical Complex.

3. Oxidation of Chloride Ion and Subsequent Chlorination of Organic Compounds by

Oxoiron(IV) Porphyrin π-Cation Radical Complexes3) Chloroperoxidase (CPO) and myeloperoxidase (MPO) are the only heme peroxidases that catalyze oxidation of chloride ion with hydrogen peroxide. CPO is an enzyme of Caldariomyces fumago and catalyzes chlorination reactions in the biosynthesis of chlorine-containing compounds. CPO is also known to exhibit peroxidase, catalase, and cytochrome P450-like activities. CPO has a thiolate heme axial ligand like cytochrome P450. This makes CPO distinct from other heme peroxidases which have a histidine imidazole as the heme axial ligand. On the other hand, MPO is found in the granules of myelocytes (precursors of neutrophils), and works as a major component of the antimicrobial system of neutrophils. MPO belongs to the animal peroxidase superfamily and has an imidazole heme axial ligand.

Numerous biological studies have suggested that an oxoiron(IV) porphyrin π-cation radical species known as compound I, is responsible for oxidation of chloride ion and for addition of a chloride ion to the ferryl oxygen atom of compound I to produce a transient ferric hypochlorite complex, Fe(III)-OCl. The ferric hypochlorite complex is believed to act as a key compound in the reactions leading to chlorination of organic substrates by CPO and antimicrobial activity in MPO. Although the oxidation process has been studied by multi-mixing stopped-flow experiments in which transiently formed compound I were reacted with chloride ion, spectroscopic evidence for the formation of the ferric hypochlorite complex has not been obtained and it remains unclear how compound I oxidizes chloride ion. Furthermore, the identity of the true chlorinating agent in the subsequent chlorination of organic substrates is not known and more information is needed about the exact roles of the hypochlorite adduct, free hypochlorous acid, and Cl_2. Here, we report the direct observation of oxidation of chloride ion with synthetic compound I model complexes and subsequent reactions leading to chlorination of organic compounds.

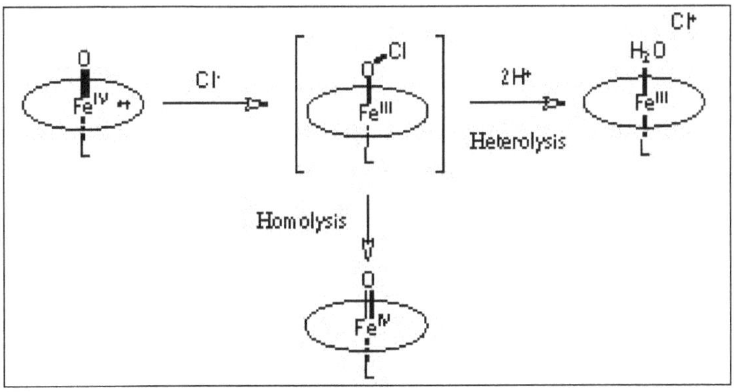

Cyclic Voltammogram of Oxoiron(IV) Porphyrin π-Cation Radical Complex.

Design of Functional Metalloproteins

Metalloprotein Design in *De Novo* Designed Scaffolds

The purest and most challenging form of metalloprotein design is *de novo* design or design "from scratch". *De novo* design of a metalloprotein involves constructing a polypeptide sequence not directly related any natural protein, to precisely fold into a defined three-dimensional structure that binds a metal ion. Most work in this area has focused on introducing metal-binding sites into designed α-helical bundles, which are among the first *de novo* designed proteins. These α-helical bundles are a common scaffold for a number of heme proteins. Therefore, not surprisingly, heme centers have been among the first and most common metal centers incorporated into *de novo* α-helical bundles, from mono heme binding to four heme maquettes mimicking long-range electron transfer proteins. These proteins were rationally designed by direct comparison to native proteins, such as cytochrome $bc_{1,}$ for clues as to the placement of essential heme binding and structure-stabilizing residues. In addition to electron transfer and proton-coupled exchange functions, a functional aniline hydroxylase, peroxidase, and oxygenase have also been engineered into *de novo* proteins. An exciting development in this field is the design of a di-heme containing four-helix bundle with oxygen transport properties, whose O_2 affinities and exchange timescales match those of natural globins. Remarkably, this designed protein binds O_2 even tighter than CO. A key to the success is reduction of helical-interface mobility to exclude water from the heme binding site, thereby reducing heme oxidation and increasing O_2 binding stability.

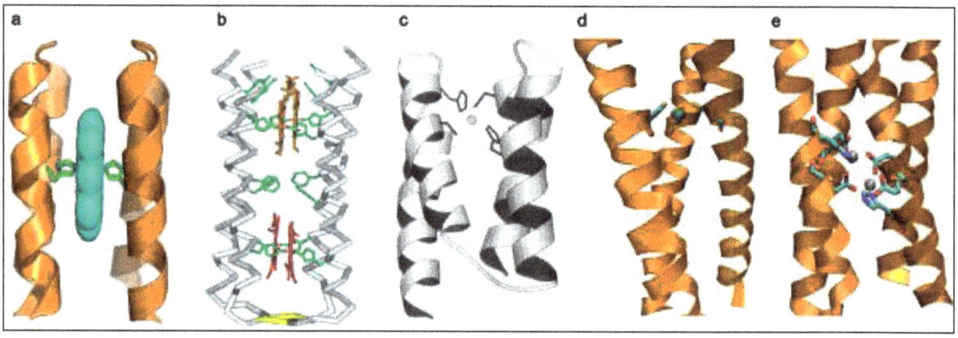

Designed metalloproteins using *de novo* designed scaffolds.

In figure above, a, Computer model of a bis-His ligated mono-heme α-helical bundle. b, Computer model of a bis-His ligated multi-heme four α-helical bundle. c, Computer model of Zn(II) bound His_2Cys_2 motif of a Zn-finger mimic. d, X-ray crystal structure of As(III) bound three-stranded coiled-coil (PDB 2JGO). e, X-ray crystal structure of di-Zn(II) Due Ferro 1; PDB 1EC5).

In addition to heme, other metal ions/cofactors have been engineered into de novo designed helical bundles by introducing metal-binding ligands at specific locations to mimic those in native proteins. Examples include the His_2Cys_2 ligand set found in Zn-finger proteins and the His_3 set in carbonic anhydrase15 The x-ray crystal structure of an arsenic bound three-stranded coiled-coil protein, resembling the putative tris Cys metal-binding environment of the natural arsenic binding protein, ArsR, has been reported, affording the first biologically relevant small molecule mimic of an AsS_3 coordination16. Metal-binding ligands Glu and His have also been introduced by retrostructural analysis, whereby the active site of a protein can be separated into idealized secondary structural components (e.g., an α-helix and a β-hairpin), and mathematically described as a root-mean-square deviation from the idealized structure. This strategy led to the construction of dinuclear metalloproteins in α-helical bundles that can bind two Zn(II), Co(II), or Fe(II), with the Zn(II) crystal structure shown to be structurally analogous to the diiron site of ribonuclease reductase and ferritin. One of these diiron proteins was reported to catalyze the two electron oxidation of 4-aminophenol, similar to the reaction catalyzed by diiron phenol oxidases.

Taking a cue from naturally occurring Cys rich coordination sites in proteins, the introduction of Cys ligands led to the development of α-helical coiled-coil metalloproteins capable of binding Cu(II), Zn(II), Cd(II), Hg(II), and As(III). One exciting aspect of de novo design is the demonstration that these de novo proteins can stabilize unusual metal coordination states, such as three-coordinate Hg(II) by a tri-helical bundle over the bis-coordination often preferred by Hg(II). In other cases, metal-binding can actually direct the folding and assembly of an α-helical bundle from either a random coil, less folded, or misfolded state, as was the case of Cd(II) and Hg(II) binding to Cys containing peptides predisposed to form helices. Interestingly, geometry-selective binding of two different Cd(II) in one designed protein has been achieved and confirmed by the correlation of ^{113}Cd NMR and ^{111m}Cd Perturbed Angular Correlation spectroscopy.

In stark contrast to the design of α-helical proteins, de novo design of β-structure proteins is still in its infancy, because there were few model systems with which to study β-sheets in isolation from other protein structures. The de novo design of a redox active rubredoxin mimic, RM1, is a rare example of a structural and functional metallo-β-sheet protein. RM1 was shown to bind iron and reversibly cycle between the Fe(II/III) oxidation states, mimicking native rubredoxin. The RM1 β-hairpin structure was designed computationally by constraining the positions of the peptide that were strategic in mimicking and/or stabilizing the β-structure and subsequent use of another computer program, SCADS to select the most probable amino acids for the remaining positions.

Metalloprotein Design in Native Protein Scaffolds

While designing metalloproteins using de novo designed scaffolds offers the prospect of complete control over a protein's structure and metal-binding properties, our current knowledge of protein folding limits the number of de novo designed scaffolds to only a few types, such as the α-helical bundles discussed above. It has been observed that β-sheet containing proteins tend to support more rigid and more pre-organized metal-binding sites than α-helical proteins, which are inherently more flexible. Unlike de novo designed scaffolds, the Protein Data Bank contains ~1000 natural protein scaffold types, most of which maintain the same fold and similar stability even after numerous mutations. Therefore, metalloprotein design using native protein scaffolds provides many more scaffold choices that are more tolerant to mutations, allowing for the incorporation of metal-binding sites into proteins with less concern over decreased protein stability.

When chosen carefully, native proteins are also relatively easier to crystallize than de novo designed scaffolds, making 3D characterization possible, an objective that is critical to success. Perhaps the strongest argument for designing metalloproteins using native scaffolds is that Nature employs the same approach since the same scaffold is often seen in numerous proteins with diverse metal-binding motifs and functions, suggesting that these native scaffolds are robust and modifiable. For example, the Greek key β-barrel fold has been shown to be used by ~600 different types of proteins with diverse functions, ranging from oxidase, reductase, amylase, and dismutase activities, for instance. Unearthing the secret of how Nature is able to use the same scaffold to design a variety of metal-binding sites is an important goal of metalloprotein design.

Biochemical techniques, such as site-directed mutagenesis, have been used extensively to study the function of metalloproteins. The loss of function accompanied by certain mutations (usually to highly conserved residues) allows identification of residues essential for function. Although serving a different purpose, the same mutagenesis techniques can be used for metalloprotein design to impart new function into a protein scaffold by the introduction of residues that bind metal ions. This approach can be in the form of redesigning existing metal-binding sites to introduce novel functionality, introducing mononuclear metal-binding sites to proteins that do not natively bind metal ions, or by introducing homonuclear or heteronuclear metal-binding sites into proteins. The designs are often aided by empirical approaches through knowledge/experience, rational design using computer programs, or combinatorial selections.

The redesign of an existing metal-binding site to introduce novel function or metal specificity is the simplest form of metalloprotein design, as this approach relies on the structural differences between template and target proteins, and is amenable to empirical designs through knowledge/experience. Despite its simple form, this approach is still quite powerful in elucidating structural features important for the change, or gain of function by redesign. Heme proteins have been extensively redesigned in this way. They are one of the most diverse classes of proteins, with functions ranging from electron transfer, small molecule (e.g., oxygen and NO) transport, sensing, and activation. By systematically changing these characteristics of heme proteins, researchers have been able to convert one type of heme protein into another, as well as introduce novel function or substrate specificity into a heme protein.

The selectivity of designed metalloproteins for specific metals has also been utilized for the purpose of metal-sensing applications. Ratiometric Zn(II) sensors have been created by designing Zn(II) binding sites into green fluorescent proteins and by taking advantage of the resulting fluorescent signal variations due to Zn(II)-dependent conformational changes. The metal-binding site of NikR, a DNA binding protein, has been redesigned to bind UO_2^{2+} rather than its native Ni^{2+} cofactor, resulting in an artificial protein that binds DNA only in the presence of UO22+. In addition to sensors, designed metal-binding sites have allowed modulation of protein-protein interactions, which are at the heart of many diverse biological functions. Because of complicated non-covalent interactions, it has been difficult to control the interactions well. These designed metal-binding sites allowed specific control of these interactions.

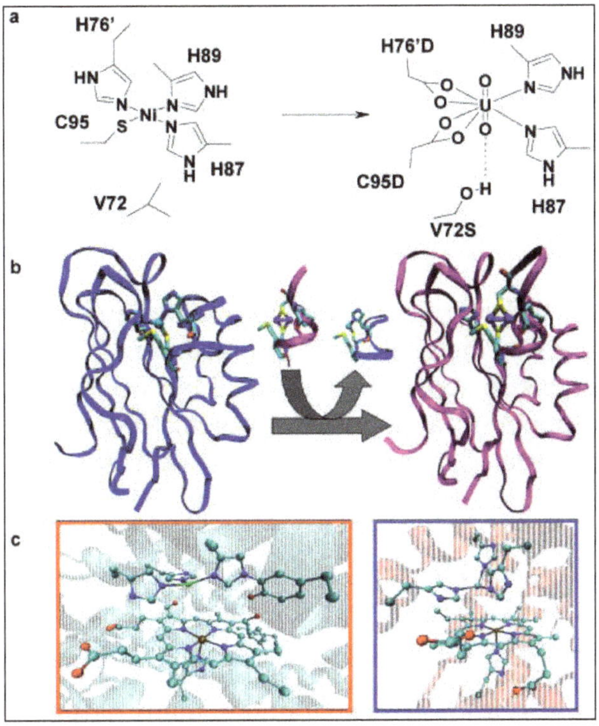

Designed metalloproteins using native scaffolds.

In figure, a, Native nickel binding site of NikR (left) and the re-engineered UO_2^{2+} binding site of the mutated NikR (right). b, Loop directed mutagenesis of blue copper azurin to yield the dinuclear, purple CuA azurin construct. c, Catalytic heme-copper center in C_cO (left) and the designed heme-cooper model in sperm whale Mb (right).

More complex metal-binding site redesign may require rational design using computational biology, such as the simultaneous incorporation and adjustment of functional elements to convert the GlyII enzyme from a glyoxalase, which hydrolyzes thioester bonds, to a β-lactamase. In order to impart the new functionality, the metal-binding site had to be significantly redesigned in order to change the metal-binding geometry, as well as the substrate binding pocket.

While redesign of metal-binding sites can offer insight into the different structural features between the template and target proteins, those structural features common to the metal-binding sites of both proteins will remain hidden. Introducing metal-binding sites into a protein location where no native metal-binding site is found, can overcome this limitation, although this approach requires a higher level of complexity. A common strategy is to base the design on structural homology between the template protein which contains no metal ions and the target metalloprotein, to introduce metal-binding sites at the corresponding positions. Using this strategy, new Zn(II) binding sites have been introduced into charybdotoxin and retinal binding protein to mimic carbonic anhydrase (CA).

When there is no structural homology, computer aided design using programs such as Metal-Search and Dezymer, has become indispensable. A Ca(II), Pb(II), and $[Fe(Cys)_4]$ rubredoxin center have all been designed into scaffold proteins using computer aided design. In the case of an artificial rubredoxin designed into the non-metalloprotein thioredoxin, it was found to undergo several rounds of air oxidation and reduction by β-mercaptoethanol, similar to native rubredoxins.

An even higher level of difficulty in metalloenzyme design is the introduction of dinuclear metal-binding sites or metal clusters into proteins. An example is the binuclear Cu_A center, consisting of two copper ions bridged by two cysteine thiolates, with each copper ion also coordinated to a histidine nitrogen. New CuA centers were designed into the cupredoxins azurin (Az) and amicyanin using a technique called loop directed mutagenesis (LDM), in which the copper binding loop of cupredoxins is replaced by the corresponding loop in C_cO that share similar structural homology. LDM has also been used to introduce lanthanide binding sites into a helix-turn-helix motif by drafting a calcium binding EF-hand loop, and such a designed protein has been shown to promote DNA and phosphate hydrolysis.

Heteronuclear bimetallic centers are also found in many natural proteins and they perform important biological functions such as oxygen reduction, lignin degradation and nitric oxide reduction. Designing metalloproteins to mimic the enzymes that carry out these complex reactions is even more challenging than designing homonuclear metal-binding sites because of the selective placement and binding of two or more different metal centers. An example is the heme-CuB center in C_cO where three histidines bind to a Cu(II) ion that is ~4.5 Å away from the heme iron. Computer-aided design was used to introduce the histidines above the heme of myoglobin (Mb) at positions corresponding to C_cO. Copper binding was confirmed by multiple spectroscopic techniques. Further studies showed the importance of the C_{uB} site in O_2 binding and reduction, as well as the significance of a proton network. This designed protein also displays interesting functional properties, such as nitric oxide reduction, similar to certain members of C_cOs. A Mn(II) binding site has also been designed into cytochrome c peroxidase, resulting in a heme-Mn(II) protein with Mn(II) oxidation activity, as in native manganese peroxidase (MnP), a protein involved in lignin biodegradation, which is a critical step in biomass conversions from plant materials to biofuels.

In addition to discrete metal-binding sites, metal nanoparticles have been introduced into proteins, including Pt nanoparticles in ferritin and a small heat shock protein. The former catalyzes size-selective olefin hydrogenation while the latter catalyzes hydrogen production.

Even though empirical approaches through knowledge/experience and rational design using computer programs have been quite successful in designing metalloproteins, it is still difficult to design certain metalloproteins, especially functional ones, due to the large number of potential interactions within the active site of a metalloenzyme. In this case, combinatorial design and directed evolution of novel metalloenzymes is another powerful choice. For example, directed evolution has been extensively applied to antibodies to develop novel metalloenzymes. Utilizing phage display, the 7G12 antibody protein was evolved to bind heme, with which the oxidation of tyramine could be catalyzed. A similar technique has recently been expanded to find antibodies for inorganic ligands, such as BINAP and BINOL, with the goal to develop completely novel metalloenzymes. A striking example of this approach is the use of directed evolution to obtain a new cyt P450 with the ability to selectively hydroxylate propane, a function that has not yet been achieved by small molecule catalysts.

Designing Novel Metalloproteins Containing Unnatural Amino Acids or Non-Native Metal Cofactors

Metalloprotein design using de novo designed or native protein scaffolds has succeeded in providing valuable insights into the fundamental rules that govern protein structure and function. Replicating native protein structure and function, although quite challenging, is not the only goal of protein designers; they want build proteins with properties beyond those found in Nature. In recent years, the repertoire of functional groups such as amino acids or cofactors available to protein designers has been expanded beyond those available in biological systems.

A myriad of techniques have been developed for the incorporation of unnatural amino acids into a protein scaffold, each with its own advantages and limitations. Total synthesis of proteins using solid phase peptide synthesis enables incorporation of any synthetic amino acid for which successful protection/deprotection techniques can be applied, but cost and size constraints limits the use of total synthesis to smaller proteins (~60–100 amino acids). Native chemical ligation (NCL), involving covalent attachment of two synthetic peptides, has expanded this size regime, but cost is still an issue, at least for large proteins. Expressed protein ligation (EPL), the coupling of a bacterially expressed peptide fragment with a synthetic peptide fragment containing unnatural amino acids, has reduced cost and extended the protein size even further, but it often requires a cysteine residue at the point of ligation to maximize ligation efficiency, although methods have been developed to use other residues at the point of ligation, with less efficiency.

Methods not restricted by protein size include auxotroph generation, chemical residue modification, and cavity complementation, but these methods have limited versatility in terms of the variety of unnatural amino acids that can be incorporated. A promising and versatile technique is the use of custom evolved tRNA synthetase molecules for in vitro and in vivo translation of new residues through recognition of the amber stop codon. A seminal contribution from this approach was the design of a DNA cleaving protein through site-specific genetic incorporation of the DNA cleaving agent (2,2′-bipyridin-5-yl)alanine into a DNA binding protein. Upon addition of Cu(II) or Fe(II) and reducing agent to the protein, a double stranded DNA substrate was cleaved at its consensus sequence. The variety of unnatural amino acids this technique can incorporate is higher than those of auxotroph generation, chemical residue modification, or cavity complementation, but lower than those of total synthesis, NCL or EPL.

The expansion of the amino acid regime has made significant contributions toward determining systematically the role of each amino acid in tuning the activity of the metal center toward a given reactivity, and in increasing the breadth of chemical transformations performed by proteins. For instance, replacement of heme binding His residues in de novo designed four-α-helical bundles with 4-β-(pyridyl)-L-alanine and 1-methyl-L-His has resulted in an increase in the reduction potential of the metal site for the pyridyl system and a preference for five-coordinate low-spin heme in the 1-methyl-L-His system. In another system, introduction of bulky penicillamine ligands in place of cysteines led to the creation of an open metal-binding site in a ferredoxin model peptide. This methodology resulted in the isolation of the first water-stable CdS_3 complex through subtle manipulation of steric crowding around the metal center. Furthermore substitution of L-amino acids for their D-enantiomers resulted in fine control between 3- and 4-coordinate Cd(II) complexes in a de novo designed metallopeptide. Finally, the role of a conserved Tyr residue near the active site of rubredoxin was probed via a set of unnatural Tyr analogs in which the para -OH group was replaced with -H (native Phe), -F, $-NO_2$, and -CN, spanning a range of electron withdrawing strengths. Interestingly, a linear correlation was found between the reduction potential of the nearby heme and the Hammett σ_p value, but no such trend is observed with dipole moment or residue size.

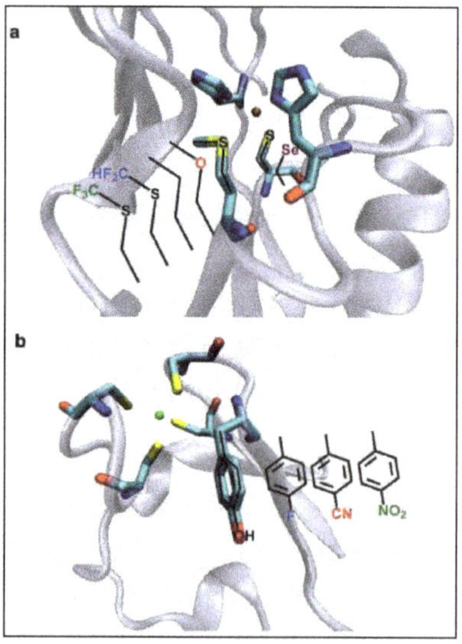

Site specific incorporation of unnatural amino acids into a protein scaffold for tuning of metal properties. Crystal structures of a, rubredoxin (PDB: 1CAA) and b, azurin (PDB: 4AZU) showing variant residue location for unnatural amino acid incorporation.

A powerful advantage to the introduction of unnatural amino acids is the ability to probe the role of the amide backbone, since site-directed mutagenesis using natural amino acids cannot modify the protein backbone. Both the carbonyl and nitrogen functionalities of the protein backbone have been implicated in critical and specific hydrogen bonding interactions, either directly to the metal, or to the metal ligands. For example, the Fe_4S_4 cluster of high potential iron proteins (HiPIPs) has such an amide interaction, which was found to be responsible for stabilizing the reduced form through attenuation of a metal bound sulfur. Replacement of this amide linkage with an ester linkage indeed resulted in a lowering of the reduction potential by ~100 mV.

Through the use of EPL, the metal bound Cys and Met residues in the electron transfer protein azurin were replaced with isostructurally similar unnatural amino acids. Replacing Cys with selenocysteine resulted in drastically different spectral features, but did little to affect the reduction potential. In contrast, replacing methionine with selenomethione and other isostructural variants resulted in spectral features similar to the wild type protein, but with drastically different reduction potentials. The use of unnatural amino acids helped to identify the hydrophobicity of the residues at this position as the major factor in tuning the reduction potential of the protein.

The repertoire of non-native functionalities has also been expanded through the incorporation of non-native metal containing cofactors into the protein scaffold. Biological systems incorporate metal cofactors such as heme through both non-covalent and covalent means. Among the covalent approach, Nature uses both single and dual point anchoring strategies to position the prosthetic group within the protein scaffold. These strategies have all been utilized for incorporation of non-native metal cofactors, with the non-covalent approach being the most extensively used strategy. For example, replacement of the native heme cofactors of myoglobin and horseradish peroxidase (HRP) with the less symmetrical structural isomer, porphycene results in enhanced peroxidase activity, highlighted by a twelve-fold increase in oxidation of thioanisole when compared to native HRP. In addition to native protein scaffolds, de novo designed α-helical bundles have also been engineered to bind non-biological heme cofactors preferentially over biological hemes.

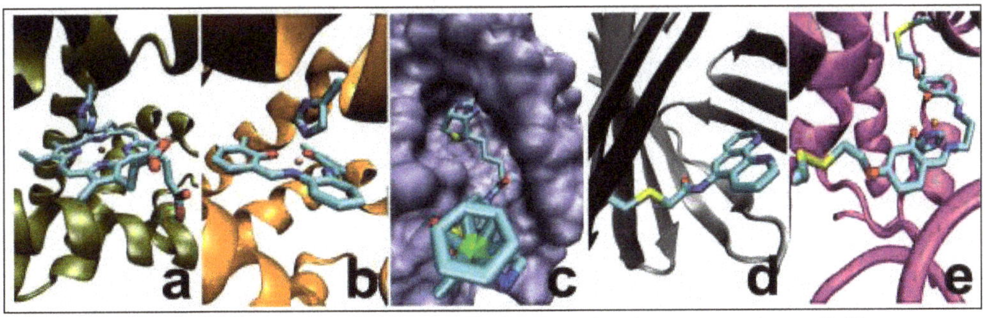

Strategies for non-native cofactor incorporation into a protein scaffold.

Native cofactor substitution of heme b in myoglobin for a, Fe-porphycene and b, 3,3-Cr salophen, exploiting the structural and dative-bonding similarities between the native cofactor and the surrogate. Affinity tagging of the catalyst with a protein-selective linker such as the biotin-streptavidin couple, c, allows for versatile attachment of non-native cofactors. Covalent strategies include a single attachment strategy as in d, showing the adipocyte lipid binding protein (ALBP) linked through Cys117 to a phenanthrolene complex; and dual covalent attachment as in e, showing dual anchoring of a Mn-salen complex into the myoglobin scaffold. Crystal structure of Fe porphycene, Cr(3,3′-Me_2-salophen)-Mb complex, biotin-streptavidin complex, and phenanthroline-ALBP (PDB codes: 2D6C, 1J3F, 2QCB, and 1A18, respectively). Computer model of Mb(L72C/Y103C) generated from an overlay with native heme b structure.

In addition to heme analogs, other planar moieties similar to heme have also been substituted into heme proteins. For example, strong binding of Mn, Fe, and Cr Schiff base complexes into a myoglobin scaffold was observed after careful design of the protein to relieve steric repulsion in the active site. Such rational design has resulted in novel asymmetric catalysts capable of oxidation rate enhancement and chiral induction.

To incorporate non-planar metal-containing cofactors into proteins where there is no native binding site is much more challenging. One way to meet such a challenge is through derivatization of the catalyst with a moiety exhibiting natural bio-affinity for a protein target, through, for example, the use of the high-affinity biotin-streptavidin couple ($K_a = 1 \times 10^{15}\ M^{-1}$). In this way, metal cofactors have been incorporated into the streptavidin scaffold through covalent attachment to a biotin moietyresulting in artificial biocatalysts with new activities including hydrogenation of alkenes, transfer-hydrogenation of ketones, and C-C bond formation through allylic alkylation, and up to 99% ee.

While the use of biotin and avidin/streptavidin to incorporate non-native metal cofactors is very powerful, not all cofactors are amenable to such a strategy, as binding affinity and location may not be readily controlled. Covalent attachment is an effective and versatile approach to attach a cofactor to the protein at a defined location through selective conjugation techniques involving engineered or native Cys or Lys residues. This strategy has been utilized to attach DNA cleaving agents such as 1,10-phenanthroline-copper and iron-EDTA to DNA binding proteins for the generation of artificial nucleases,and 1,10-phenanthroline-copper for enantioselective hydrolysis.

While the single-point covalent approach makes it possible to attach a non-native cofactor at a defined location, it does not necessarily fix the conformation of the cofactor inside the protein. For example, single covalent attachment of a 4-(2-methanesulfonylthio-ethoxy)salicylidene-1,2-ethanediamino-manganese(III) (Mn-Salen) cofactor to apo myoglobin through a single cysteine-selective linker resulted in only 12% ee toward sulfoxidation of thioanisole, suggesting that the Mn-Salen conformation may be too flexible inside Mb to facilitate high enantioselectivity. Addition of a second linker to the cofactor and selective cavity positioning led to a significant increase in catalytic efficiency (over 7-fold increase) and selectivity (from 12% to 51%). Furthermore, it has been shown that such incorporation of Mn-Salen into a myoglobin scaffold enhances significantly the chemoselectivity in oxidation of thioanisole with H_2O_2, virtually eliminating over oxidation to the sulfone product through electrostatic exclusion of the sulfoxide product from the site.

Measures of Success, Lessons Learned and Future Challenges

Tremendous progress has been made in this exciting field of synthetic biology. Success is measured not only by how close the designed protein is to the native protein, but also by the insight gained from the process. For example, the crystal structure of a designed Cu_A site is almost identical to that of native Cu_A, including both primary and secondary coordination, indicating that metalloprotein design can lead to structures closely resembling those in Nature. More importantly, two metal-binding sites, native blue copper and the designed Cu_A center, were placed in the same azurin scaffold, allowing the electron transfer properties of two different metal centers to be compared in the same protein, a task that is difficult, if not impossible to do by studying native proteins. Such a comparison provided deeper insight that the Cu_A center is a more efficient ET center than blue copper.

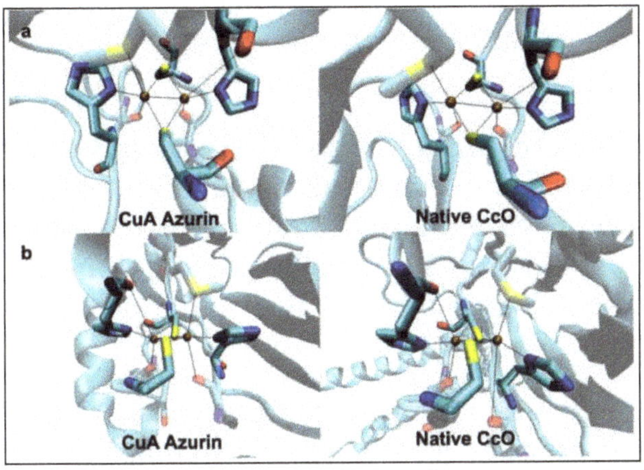

Close match between a designed metalloprotein and its native target protein. Crystal structures of a designed Cu_A site in azurin (PDB code 1CC3) and a CuA site in native CcO (PDB code: 1AR1) as viewed from above the $Cu_2(S_{cys})_2$ plane (a), and along the $Cu_2(S_{cys})_2$ plane.

Given the initial successes, it is a good time to summarize the lessons learned and to recognize future challenges. Essential to success is the recognition that structural biology must play a major role in the design process. Some early metalloprotein designs have generated much excitement in the field, only to be shown by x-ray crystallography to contain a different metal-binding site than originally designed. While spectroscopy is invaluable in providing support for the design, more concrete evidence from 3D structural characterization is essential. In addition to confirming the initial design, it provides detailed information on what has or has not been designed correctly, thus providing insights on the next level of design.

Furthermore, primary coordination sphere design is not enough in many cases, to confer structure and/or function. Consideration of the environment surrounding the primary coordination sphere is critical to successful design. One primary example is water inside the protein, where the associated hydrogen bonding interactions play essential roles in many metalloproteins, including Cyt P450, heme oxygenase (HO), and CcO. For example, even though a CuB center was designed into myoglobin that structurally mimics the heme-copper center in CcO, its function mimics HO rather than CcO when O_2 is used as an oxidant. When H_2O_2 is used as an oxidant, a ferryl species, a putative intermediate in CcO activity, was observed. These results suggest that the extra protons introduced through H_2O_2 are important for function. The introduction of an additional hydrogen bonding donor through the replacement of the b-type heme in Mb with an o-type heme, resembling the heme found in CcO, allowed suppression of HO activity and promotion of CcO activity.

Metalloprotein design has focused on the positive design of structural features responsible for correct metal-site geometry and function. However, negative design to prevent competing metal-site geometries and unwanted functions is also critical. For example, in designing dinuclear metalloproteins in a helical bundle, there was a possibility that undesirable heterotetrameric motifs and homooligomers would form since the protein was comprised of different peptides. To circumvent this problem, a computational algorithm was included to destabilize these interactions.

The metalloprotein design often results in lower stability because of changes made to the protein scaffold. Sometimes, the stability is so low that even if the design is valid, not enough stable protein can be obtained to prove the design. The stability of the protein must be improved, even if function may not be improved directly. A beautiful example is demonstrated in cytochrome P450 where only after the protein was stabilized, could the highly destabilizing mutations introduced confer novel activity.

The majority of the successes in the metalloprotein design field are derived from the construction of a geometrically correct, sterically compatible primary coordination sphere sufficient to reproduce the structure and function of a desired metal center, such as a bis-His heme or His_2Cys_2 tetrahedral Zn(II) site. However, many metal-binding sites in proteins, such as the type 1 copper and CuA centers, are not necessarily in their preferred metal ion geometric states. Therefore, one future challenge is to design metal sites and metal clusters with unique geometric requirements. Other future challenges include the design of metal sites at the interface of two or more proteins, which occur often in proteins but have not been designed until recently, and the design of metal-binding sites from membrane proteins, which has also been rare. An even higher level of challenge is the design of protein metal-binding sites requiring helpers called chaperones. Consideration of the interactions between metal ions and protein hosts is not enough to confer function, making their interactions with chaperones necessary in the design process. In addition to metal-binding site design, functional metalloproteins also require the design of a substrate-binding site for catalysis. An even higher level of complexity includes coupling the redox reaction of a designed metal-binding site to proton transfer, conformational change, or charge separation.

To meet these challenges, advances in a number of fields are required. The most important requirement is the development of better computer programs for designing metal-binding sites in proteins, including consideration of structural features beyond the primary coordination sphere. In contrast to well-defined force fields for peptide bonds and amino acid side chains, most force fields for metal-binding sites are ill-defined because there are many more variables involved in defining a metal-binding site. These variables include variations on the identity of the metal ion, different oxidation states of the same metal ions, and different ligand and geometric preferences of different metal ions. Density functional theory has been excellent in defining small metal complexes. Extending its success to metal-binding sites in proteins, together with other leading algorithms, such as QM/MM (quantum mechanics/molecular mechanics), is required for the successful rational design of metalloproteins. Despite these challenges, metalloprotein design will remain a vibrant field where exciting discoveries of new insights and novel enzymes will occur every day.

Iron Sulphur Proteins

The understanding of structures and functions of iron sulfur proteins is an area of bio-inorganic chemistry which has developed into a subject of great significance over the last two decades. These groups of non-heme iron-sulfur (Fe-S) compounds is involved in electron transfer reactions in biological systems and are thus ubiquitous in nature. They are thought to be fundamental to such processes as photosynthesis, nitrogen fixation and various metabolic reactions. Yet the precise nature of their composition and stereochemistry was not established until very recently. In fact, it was only as the result (unintentionally, as all significant discoveries tend to be) of advances made in the field of anaerobic fermentative metabolism and photosynthesis during the period 1952-63 that these Fe-S compounds were first recognised.

The main reason for such a relatively common and important class of compounds remaining undiscovered for so long is that the Fe-S redox centres, which form the active sites of these proteins, could not be extracted from the proteins until recently, and examined chemically. All the previous attempts to do so had caused the disintegration of the active site of the protein and this resulted in the complete loss of biological activity.

Optical absorption spectroscopy produces weak, broad spectra from Fe-S proteins. So the method which finally became appropriate to their detection was low temperature electron paramagnetic resonance spectroscopy. It was only after X-ray crystallographic results became available during 1968-71 that the exact composition and stereochemistry of Fe-S protein active sites were established.

Heme and Non-Heme Proteins

'Heme' is an iron chelate of a substituted porphyrin and is a non-protein part of a holo-enzyme. Iron could be present as FeUI) or Fe(JII). For example 'heme' is the prosthetic group in haemoglobin, myoglobin, and cytochrome. 'Non-heme' proteins are thus the ones which do not contain Fe^{ll}/Fe^{lli} chelate of porphyrin as a non-protein in a haloenzyme.

Types of Fe-S Protein Centres

Iron sulfur proteins are found to be of low molecular weight (6000-20,000) and contain iron, cysteinyl sulfurs (coming from the amino acid cysteine in the protein chain) and 'inorganic sulfur', the so called non-cysteinyl sulfur atoms. The sulfur ligands are arranged tetrahedrally about the iron atoms. The presence of inorganic sulfur is indicated through the release of H_2S gas when these proteins are treated with a mineral acid. The cluster of iron and inorganic sulfur atoms in the form $Fe_nS_{n'}$, or (nFe-n'S) where nand n' are 1,0; 2,2; 3,4; 4,4 respectively constitutes the core group in these proteins and are their redox centres. These redox centres are wrapped in the protein chain with the cysteinyl sulfurs of the protein coordinating to the irons of the core group. These proteins may be categorised into five accepted classes of Fe-S centres, according to the numbers of iron and inorganic sulfur atoms in the cluster.

Classification of Iron-Sulfur Proteins

- Rubredoxins (Fe-oS): This protein is of bacterial origin and was originally isolated from Clostridium pasteurianum but has since been found in a number of anaerobic micro-organisms. All have one iron atom which is tetrahedrally coordinated to four sulfur atoms from the four cysteine residues present in the protein chain of 55 amino acids. The redox centre is made up of only high spin Fe(III) and there are no inorganic sulfur atoms in the core group.

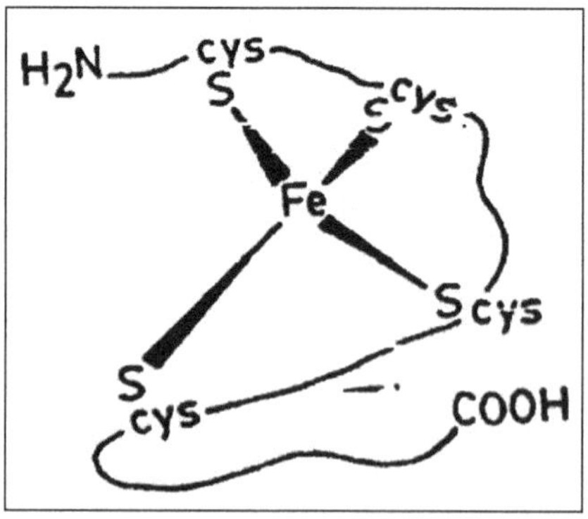

- Two-Iron Ferredoxins [2Fe-2S]: A number of these proteins (Fd) have been isolated from plant and non-plant sources. Unlike in rubredoxin, the redox centre is a cluster of equal numbers of iron and inorganic sulfur atoms represented as (2Fe-2S). This redox centre is wrapped around the protein chain, through the coordination of iron atom with the cysteinyl sulfurs. The structure of the (2Fe-2S) proteins can best be regarded as a bitetrahedral one, with the iron atoms bridged by two sulfide groups and bound additionally by two sulfurs of the cysteine groups per iron atom as represented schematically alongside. Since the iron centres are four coordinate, the crystalfield splitting is weak and thus the metals are in high spin state in both oxidised and reduced forms. One major class of (2Fe-2S) proteins is the chloroplast ferredoxin, which as the name suggests is fundamental in the photosynthetic process in plants.

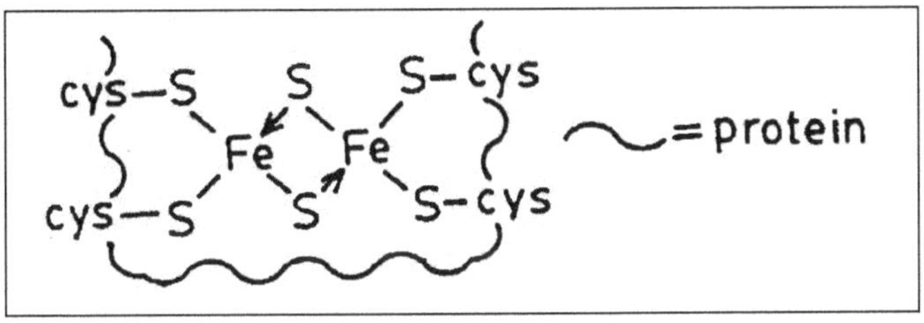

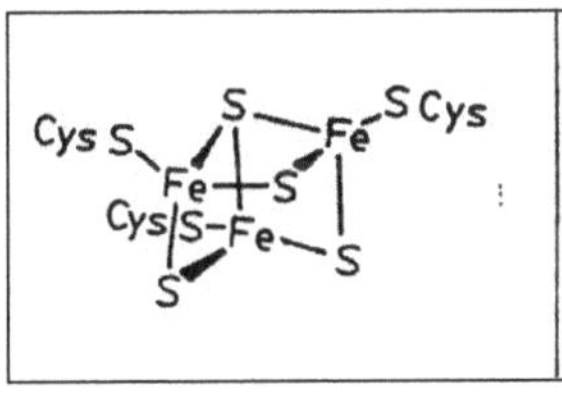

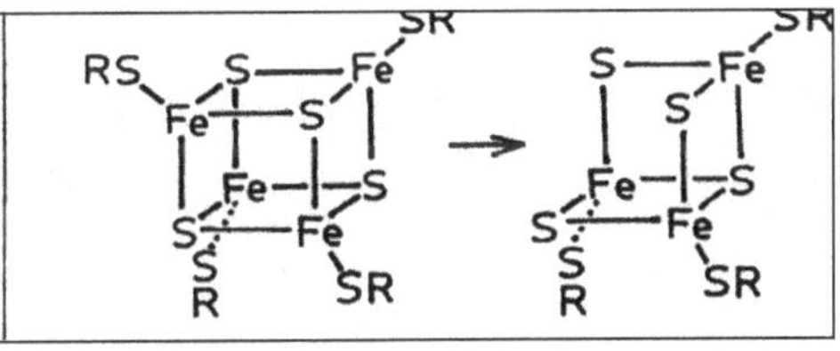

- Three-Iron Ferredoxin [3Fe-4S]: Originally the X-ray crystal structure of the 7-iron ferredoxin from Azotobacter vinelandii was interpreted as having a (4Fe-4S) and (3Fe-3S) cluster. However, an error was detected in the analysis of this structure and the tri-iron cluster was corrected as having a non-planar Fe_3S_4 geometry. This structure can be derived from (4Fe-4S) core by removing an Fe-S (cysteine) unit from one of the eight corners of the cube; a formal transformation is illustrated above. This cluster composition has been supported from EXAFS data on the A. vinelandii Fd, 3Fe-4S Fd from DesuLJovibro gigas. Their behaviour is at present poorly understood and the biological significance is still being debated.

- Four-Iron Ferredoxins (4Fe-4S): This is the most ubiquitous iron-sulfur cluster in biology and has both plant and bacterial origins. The cluster contains (4Fe-4S) as the redox centre. The iron and sulfur atoms are arranged in a cubane structure, with each iron linked to the protein via a cysteinyl sulfur. The result is two interpenetrating concentric tetrahedra of four iron and four sulfide atoms with mean Fe-Fe and S-S distances being 2.7SA and 3.ssA respectively. It may be seen that the structure can be considered as a pair of interacting Fe_2S_2 groups.

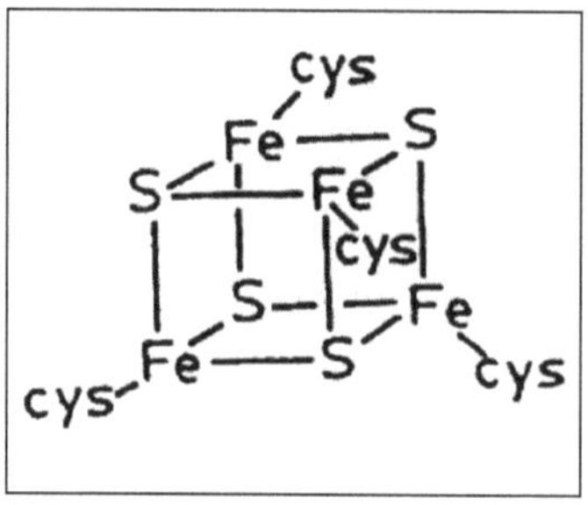

- Eight-Iron Ferredoxin [8Fe-8S]: It is of bacterial origin (Peptococcus aerogenes) and contains two Fe_4S_4 cubes separated from each other by 12A.

Synthetic Models of Iron-Sulfur Protein Centres

A major part of the work leading to the current understanding of Fe-S centres in natural proteins has been the synthesis and subsequent study of synthetic analogues. Synthesis of a 4FeFerredoxin centre was first achieved by Holm and his group. Treatment of a methanolic solution of iron(III) chloride and sodium hydrogen sulfide with three equivalents of $Na(SCH_2Ph)$ results in an intensely coloured solution of $[Fe_4S_4(SCH_2Ph)_4]^{-2}$ cluster. In this reaction some of the thiolate ligand is consumed to reduce half of the iron formally to the + 2 oxidation level required for the $[4Fe\text{-}4S]^{2+}$ unit.

$$4FeCl_3 + 12RS^- \longrightarrow 4Fe(SR)_3 + 12Cl^-$$

$$\xrightarrow[4HS^-]{4OMe^-} [Fe_4S_4(SR)_4]^{2-} + RSSR + 6RS^- + 4MeOH$$

Alternatively and in slight modification to this synthetic route, Garner and his group used $FeCI_3/FeCI_2$ together with elemental sulfur instead ofNaHS in the presence of thiolate. The thiolate serves as a reducing agent to yield the tetranuclear species in a 'self-assembly' reaction:

$$4FeCl_3 / FeCl_2 + 4S + 14[SR]^- \longrightarrow \left[Fe_4S_4(SR)_4\right]^{2-} + 12Cl^- + 5RSSR$$

Synthetic routes to the analogous [2Fe-2S] family were discovered shortly after the preparation of tetranuclear cluster, and the first characterisation of mononuclear rubredoxin models subsequently followed. Typically in these model systems ligation of the iron centre(s) by simple alkyl- or aryl-thio groups replaces that of the cysteine residues of the protein polypeptide.

Behaviour of Synthetic Fe-S Centres and Core Extrusion from Natural Proteins

The study of properties of core clusters in proteins was greatly helped by the successful synthesis of their isostructural analogues. The investigations of the core cluster in natural proteins were greatly hampered because of long polypeptide chains into which these were embedded.

One reaction that proved to be of practical significance is ligand substitution in synthetic core clusters in which the terminal thiolates are replaced in a stepwise fashion.

$$\left[Fe_4S_4(SR)_4\right]^{2-} + nR'SH <---> \left[Fe_4S_4(SR)_{4-n}(SR')_n\right]^{2-} + nRSH$$

with the equilibrium favoured to the right if R'SH is more acidic than RSH. This facile thiol exchange together with the observation that the low molecular weight iron - sulfur proteins can be reversibly unfolded by solvents such as dimethyl sulfoxide, led to the idea that it should be possible to extrude iron-sulfur cores from natural proteins using an appropriate thiol. This approach has been applied successfully to distinguish 2Fe- and 4Fe-ferrodoxin centres in natural systems.

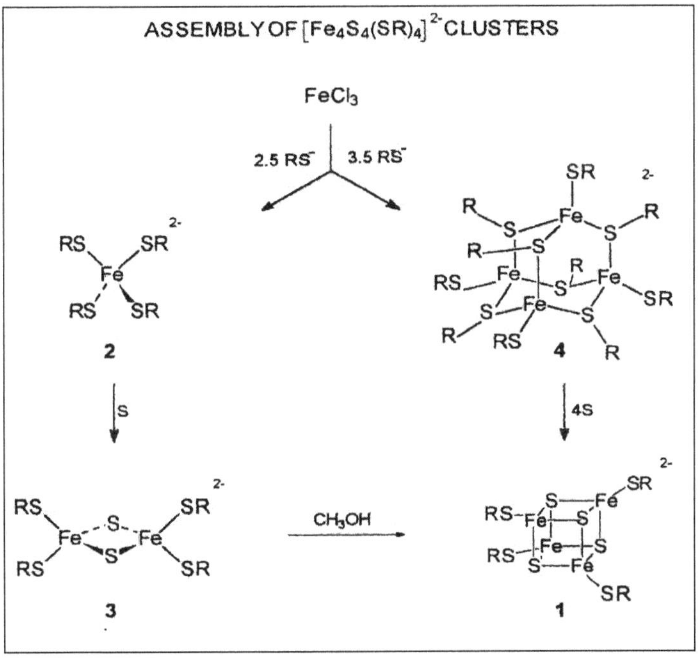

Depiction of the course of reactions resulting in assembly of $Fe_4S_4(SR)_4^{2-}$ clusters (1) via the intermediates $Fe_2S_2(SR)_4^{2-}$(3) and $Fe_4(SR)_{10}^{2-}$ (4).

Interchange Between Natural and Synthetic Clusters

Orme-Johnson has evolved parallel reactions with proteins in which the iron-sulfur core of one haloprotein was extracted and bound to another apoprotein, the iron-sulfur core of the latter having been removed by acidification (I). The reconstitution of active C. pasteurianumfrom the apoprotein and the synthetic $[Fe_4S_4(SCH_2CH_2oH_4)^{2-}$ demonstrates the complete interchange between natural and synthetic clusters (II).

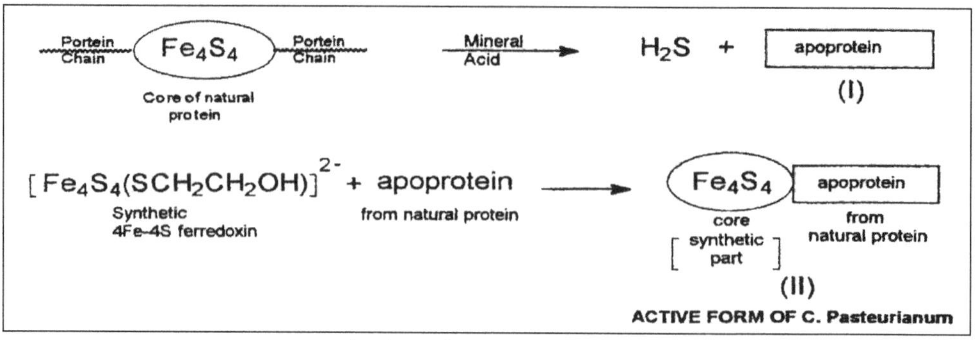

Reconstitution of active C. pasteurianum.

These experiments show that the redox core in both the synthetic and natural clusters are isoelectronic and isodimensional.

High Potential Iron Proteins (HiPIP)

A confusing part of the early studies of iron-sulfur proteins was the fact that the proteins with apparently the same core (4Fe-4S) and having isostructural cubane geometry exhibited different potentials. Such proteins called HiPIP are of bacterial origin (Chromafium vinosurrn) and have redox potentials about 0.75 V, more positive than those of the normal 4-Fe ferrodoxin (Fd). It has also been established that $HIPIP_{reduced}$ is isoelectronic with $Fd_{oxidised}$ and that both are diamagnetic in these states. Though $HiPiP_{red}$. and Fd_{ox} are both isostroctural and isoelectronic, the reason for the difference in their redox behavior has not been as yet clearly understood.

Redox Properties of Natural and Synthetic Centres

One area of Fe-S proteins which is closely linked with their function is their redox behaviour. It might be useful to consider this aspect and to compare with the synthetic centres. With the exception of high potential iron protein (HiPIP), all Fe-S proteins show fairly low redox potentials and generally function as electron carriers in conjugation with some other high molecular weight protein that has enzymatic activity. With all except the 8Fe-8S proteins, it appears that just one electron is involved and the synthetic Fe-S centres show analogous behaviour. Fe-oS rubredoxin centres are not at all well investigated in terms of functions but are found in anaerobes and are definitely known to be involved in fatty acidω-hydroxylation.

2Fe-ferredoxins transfer one electron at potentials from +250 mV to -400 mV. In the reduced state, Mossbauer spectra show that one iron atom is in the high spin ferrous state and one is in the high spin ferric state. When oxidised, both iron atoms of the 2Fe-2S centre are in the ferric state. It is difficult to assign formal oxidation states to the iron atoms in the reduced state. Ferredoxins may thus be regarded as a delocalised complex of iron with sulfur ligands. The unpaired electron in reduced ferredoxin can be regarded as being localised largely on the ligands utilising the n-acceptor properties of sulfur donor atoms. Further investigations show that the oxidised/reduced-ferredoxin couple has potentials of -300 m V difference from the corresponding 2-/3-potentials of synthetic peptide complexes. This has been attributed to the well-known effect of protein structure, environment and charge distribution on the redox potentials. Nevertheless care must be taken not to take the findings of the synthetic analogue Fe-S centres too far without direct supporting evidence from natural centres.

Fe-4S proteins do not appear to be as widespread as 2Feferredoxins, but their redox chemistry seems to be a little more complex. The Fe_4S_4 cluster in the protein has aD_{2d} symmetry and the synthetic model undergoes one electron redox reaction characteristic of the cluster in natural ferredoxin. No evidence has yet been produced for localised Fe(II) and Fe(III) in synthetic 4Fe-4S clusters and it seems probable that the electrons involved in redox processes are delocalised over the whole system.

	Potential/m V (vs SHE)	Oxidised	Reduced
Rubredoxin	-40	Fe^{III}	$2Fe^{III}$
2Fe-2S	-250 to -420	$2Fe^{III}$	Fe^{II} Fe^{III}
3Fe-4S	-130 to -420	$3Fe^{III}$	$2Fe^{2.25}$ Fe^{III}
4Fe-4S	-280 to -460	$4Fe^{2.5}$	$4Fe^{2.5}$
4Fe-4S HiPIP	+280 to +360	$4Fe^{2.7}$	$4Fe^{2.5}$

4Fe-ferredoxin can attain three oxidation states (1-, 2-, 3-) as opposed to the two of Fe-oS ad 2Fe-2S centres (2-, 3-). The existence of four oxidation states (1-, 2-, 3-, 4-) by the corresponding synthetic centres shows the complexity of the redox chemistry of 4Fe-ferredoxins. Considerable discrepancy is in fact found between the potentials of the analogue reactions and those of the proteins but this is probably attributable to a combination of factors already mentioned. Important 4Fe-4S centre proteins (in so far as most information is available about them) are the 8Fe-8S and HiPIP proteins. The HiPIP is exceptional in more ways than its high redox potential- the FeS cube in HiPIP shrinks a little upon oxidation. The shape of the polyhedron in the reduced state is very close to that of the oxidised state of bacterial ferredoxin. The electronic properties of the synthetic dianion indicate that the oxidation level is equivalent to that of $HiPIP_{red}$ and Fd_{ox}, while the electronic properties of the reduced species $[Fe_4S_4(SR)_4]^{3-}$ are closely similar to those of Fd_{red} and include a g= 1.94 EPR signal. Clearly the redox properties of the synthetic model are similar to those of ferredoxin rather than HiPIP.

Finally, the two identical 4Fe-centres found in 8Fe-8S proteins can be a highly efficient arrangement in terms of electron transport since there is a high 'cluster to protein weight ratio'. As yet such centres have only been found in anaerobic bacteria and the understanding of the way in which the two discrete clusters interact in Fe-S proteins is limited. This also reflects the inconsequential progress made in the synthesis of their analogues in the laboratory. However their function in these bacteria is known to be extensive and they are thought to be involved in key metabolic processes such as hydrogen uptake, NAD reduction, ATP formation, pyruvate metabolism and nitrogen fixation.

It thus seems clear that in each type of ferredoxins, there exists at least two accessible redox states per active site. They function primarily in electron transport sequences, rather than acting as a site for substrate binding and conversion.

- EPR (Electron Paramagnetic Resonance, also referred to as Electron Spin Resonance): A spectroscopic technique by which unpaired electrons can be detected in ions and molecules.
- EXAFS (Extended X-ray Absorption Fine Structure): This is based on the back scattering of photoelectrons off ligands, and provides information on radial distances, number and types of bonded atoms.
- Mossbauer Spectroscopy (named after its discoverer): This spectroscopic technique, in which fluorescence resulting from the absorption and emission of a y-ray is observed, provides information about the electronic environment at the nucleus of an atom.
- Redox Potentials: The redox potential of a metal atom/ion/molecule essentially tells us about the ease with which the species can gain or lose electrons. The redox potentials of metal ions can be significantly attenuated by ligands. Redox potential con be measured by a variety of techniques, the most common one being Cyclic Voltammetry.

Synthetic Model Complexes of Iron-Sulphur Proteins

Rubredoxin Model Complexes

Rubredoxins are low molecular weight iron proteins containing an active site that is composed of one Fe ion tetrahedrally coordinated by four cysteine thiolate ligands. Synthetic models, of $Fe^{II}(Rd_{red})$ and $Fe^{III}(Rd_{ox})$ that can be readily obtained, have been discussed in the ensuing sections.

The common synthetic pathway for Fe(II) reduced site analogues of rubredoxin generally accessible through simple reaction in ethanol or methanol and subsequent precipitation with an appropriate cation has been demonstrated below:

$$FeCl_2 + 4RS^- \rightarrow \left[Fe(SR)_4\right]^{2-} + 2Cl^-$$

On the other hand, Fe(III) oxidized site analogues are not stable and therefore ligand exchange reactions has been widely utilized to develop stable Fe(III) model complexes with R=alkyl and aryl.

$$\left[Fe(OC_6H_3-2,6-Me_2)_4\right]- + 4RSH \rightarrow \left[Fe(SR)_4\right]^- + 4\ 2,6-M_{e2}C_6H_3OH$$

Structure of Rubredoxin Model Complexes

The structure of the $[Fe(RS)_4]$ core having a D_{2d} tetragonally symmetry,compressed along the S4 axis has been shown in Figure below. Studies have revealed that the bond lengths and angles for both typical rubredoxin and its analogues are consistent, with typical ranges of Fe^{III} –S = 2.25–2.31 A° and Fe^{II}–S = 2.29–2.39 A°. It is also evident from the obtained structural data that the distortions within the enzyme core in the synthetic model compounds are caused by the surrounding protein.

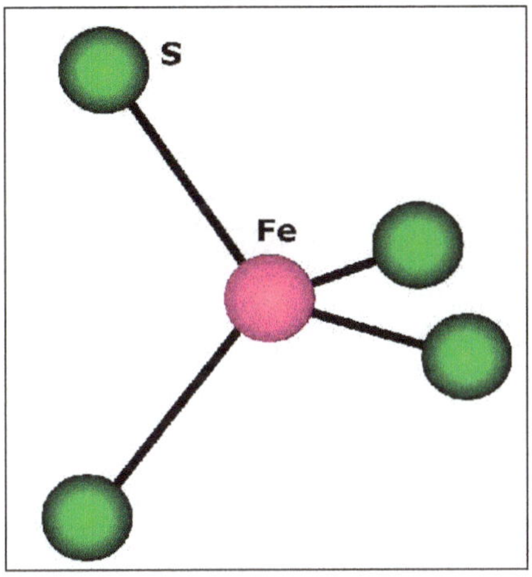

Structure of the $[Fe(RS)_4]$ core.

Properties and Biochemical Activity of Rubredoxin Model Complexes

The reduction of the Fe^{III} ion has been considered to be the most crucial redox reaction related to protein active sites and model complexes.

$$\left[Fe(SR)_4\right]^- + e^- \rightarrow \left[Fe(SR)_4\right]^{2-}$$

In case of rubredoxin clusters, redox potential not only depends on protein environment but also influenced by effects of solvent such as electrostatic interactions, pH and ionic strength. The synthetic model complex also supports the redox reaction and the potential for this couple exhibits expected shifts due to chemical environment. In addition, it is also reported that most native rubredoxin iron sulphur proteins demonstrate a reduction potential of 0.10 to −0.10 V versus SCE. On the other hand, in most cases, analogue complexes show a comparatively more negative range from −0.11 to −1.11 V.

$[Fe_2S_2]$ Model Complexes

The simplest polymetallic system, the $[Fe_2S_2]^{2+}$ cluster is found in plant-type ferredoxins as well as in Rieske proteins. In ferredoxins, the $[Fe_2S_2]^{2+}$ core binds with four thiolate ligand from Cys residues. However, the two thiolates are substituted with imidazole ligands from His residues in case of Rieske proteins. The models for [Fe_2S_2] complexes have been synthesized by series of reaction as shown below:

$$2FeCl_3 + 4RS^- + 2HS^- + 2MeO^- \rightarrow \left[Fe_2S_2(SR)_4\right]^{2-} + 6Cl^- + 2MeOH$$

$$2FeCl_3 + 8RS^- + 2S \rightarrow \left[Fe_2S_2(SR)_4\right]^{2-} + 2RSSR + 6Cl^-$$

$$2\left[Fe(SR)_4\right]^{2-} + 2S \rightarrow \left[Fe_2S_2(SR)_4\right]^{2-} + RSSR + 2RS^-$$

$$2\left[Fe(SR)_4\right]^{-} + 2S \rightarrow \left[Fe_2S_2(SR)_4\right]^{2-} + 2RSSR$$

Typical ligand substitution reactions are as follows:

$$\left[Fe_2S_2(SR)^4\right]^{2-} + 4R'SH \rightarrow \left[Fe_2S_2(SR')_4\right]^{2-} + 4RSH$$

$$\left[Fe_2S_2X_4\right]^{2-} + 4RS^- \rightarrow \left[Fe_2S_2(SR)_4\right]^{2-} + 4X^-$$

Structure of [Fe_2S_2] Model Complexes

The structure of the [Fe_2S_2] core has been shown in figure below wherein two iron ions bridged by two sulfide ions and coordinated by four cysteinyl ligands (in Fe_2S_2 ferredoxins) or by two cysteines and two histidines (in Rieske proteins). In such type of clusters, Fe_2S_2 rhomb is present which is either planar or nonplanar due to slight deviations from D_{2h} symmetry. The Fe– Fe distances in binuclear site of protein clusters are generally longer while in most of the synthetic model complexes, it fall within a small range of 2.69-2.71A°.

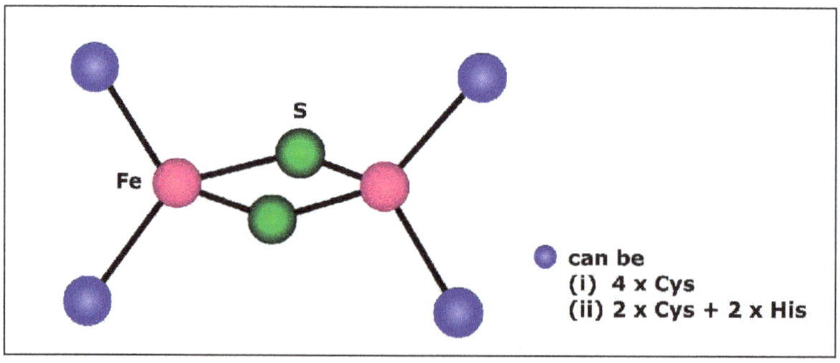

Structure of the [Fe2S_2] core.

Properties and Biochemical Activity of [Fe_2S_2] Model Complexes

Both types of iron sulphur proteins utilize $(Fe^{III})_2/Fe^{III}Fe^{II}$ redox couple in order to perform biological electron transfer reactions. The two iron atoms present in the active site of the protein are rendered inequivalent due to the surrounding protein environment even in case of the fully oxidized $(Fe^{III})_2$. However, in the synthetic analogues both iron atoms are equivalent due to the symmetry of the overall complex. It is also reported that the redox couple operates at E° = −0.42V versus SHE while in case of synthetic analogues, it works at a more negative potential. The redox properties of the synthetic clusters have been utilized for the oxidation of alcohols and keto alcohols, and also for the reduction of alkenes, alkynes and carbon dioxide.

[Fe_3S_4] Model Complexes

[Fe_3S_4]-clusters are ubiquitous iron-sulfur motifs which are predominantly found in vitro in various sources such as bacterial proteins and human iron regulatory proteins. Based on their linear or cuboidal geometry, a number of synthetic models have been synthesized. The synthesis of the linear $[Fe_3S_4(SEt)_4]^{3-}$ complex from $[Fe(SEt)_4]^{2-}$ has been described below:

$$3\left[Fe(SEt)_4\right]^{2-} + 4S \longrightarrow \left[Fe_3S_4(SEt)_4\right]^{3-} + 5/2\,EtSSEt + 3EtS^-$$

The other forms of the linear three-iron core can be generated by ligand substitution reaction as shown underneath:

$$\left[Fe_3S_4(SEt)_4\right]^{3-} + 4RSH \longrightarrow \left[Fe_3S_4(SR)_4\right]^{3-} + 4EtSH$$

The cuboidal structure of [Fe_3S_4] cluster was synthesized for the first time by Holm's group using the semirigid trifunctional cavitand 1,3,5-tris((4,6-dimethyl-3-mercaptophenyl)thio)-2,4,6- tris(p-tolylthio)benzene $(LS_3)^{3-}$ to bind with [Fe_4S_4] cubane at three of the four iron corners. The unique iron was subsequently removed by reaction with N-methylimidodiacetate.

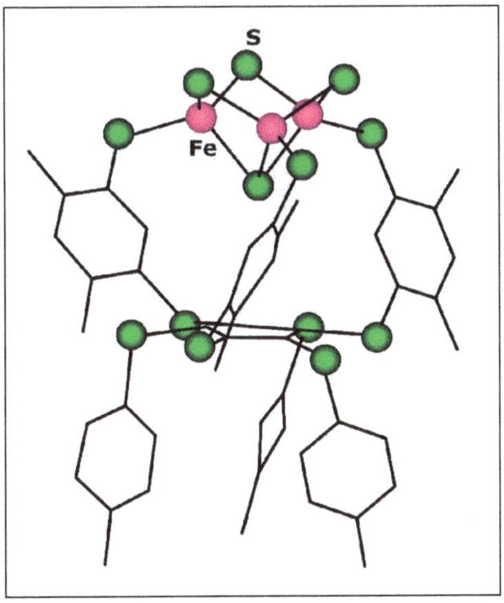

Structure of $[Fe_3S_4(LS_3)]^{3-}$.

Structures of [Fe_3S_4] Model Complexes

It is well documented from the literature that the linear cluster analogue contains two rhombs of [Fe_2S_2] core, bound to one of the iron atoms. In contrast, the structure of the cuboidal model compound is closely related to the cubane cluster (Structures of both the analogues i.e. cuboidal and linear have been depicted in Figure below). Further, it has also been found that the angles, Fe–S and Fe···Fe distances observed in synthetic analogues are in good agreement with the iron sulphur clusters.

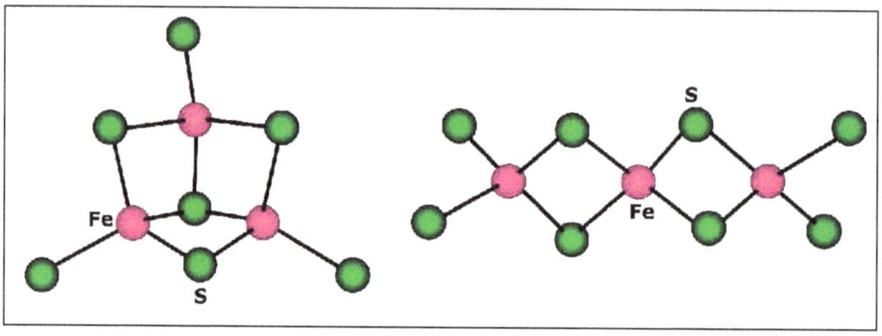

Cuboidal and linear cluster analogues of Fe_2S_2 core.

Properties and Biochemical Activity of [Fe_3S_4] Model Complexes

The three-iron in clusters are generally present either in a completely oxidized ($3Fe^{III}$) or mixed Fe^{III}/Fe^{II} oxidation states. However, their reduced forms in synthetic as well as natural systems have not been isolated till date. It is worth noting that both model complexes and naturally occurring [3Fe-4S] ferridoxins are involved in electron transfer reactions, reducing to successively more ferrous oxidation states. The oxidized forms of the synthetic models undergo reduction at negative potential value i.e. ~ −1.5 V.

Zinc

Zinc is the second most abundant trace metal in the human body, after iron. It is essential for the activity of more than 300 enzymes, covering all six classes of enzymes. Zinc in proteins can play an active catalytic role, generally as a strong lewis acid or it can act in regulatory or structural role (maintaining protein structure and stability). Metalloenzyme is an enzyme which has a bound metal ion. There are several reasons for zinc (II) being an integral component of metalloenzymes:

- The electronic configuration of zinc (II) is [Ar] $3d^{10}$. Since it has a completely filled d shell orbital, its ligand field stabilization energy is zero, irrespective of the number of coordinated ligands and hence no geometry is inherently more stable than the other. As a consequence, it readily forms low coordination complexes which have more acidity than high coordination complexes and which can easily change coordination number from 4 to 5 to 6.

- Zinc (II) being a d^{10} system has no tendency to undergo redox reactions under biological conditions.
- It is a good lewis acid.
- It rapidly exchanges ligands, allowing for rapid product dissociation.
- Unlike many other transition metals, zinc (II) does not undergo hydrolysis at low pH to form hydroxo complexes.
- Non-toxic nature of zinc.

Classification of Zinc Sites in Proteins

In zinc proteins, zinc may be present at a catalytic, cocatalytic or structural site. In all cases, removal of the bound zinc by a chelating agent can lead to the loss of enzymatic activity.

Catalytic Zinc Site

In all catalytic sites the function of zinc ion arises from its capacity to act as a lewis acid. A zinc(II) ion which has a catalytic role in the enzyme is present at the active site of the enzyme and directly participates in the bond- making or breaking step. The coordination sphere of the zinc ion is open and contains at least one water molecule along with three or four protein ligands. The typical geometry of the catalytic site is distorted tetrahedral. However, the trigonal bipyramidal geometry is also found in some enzymes.

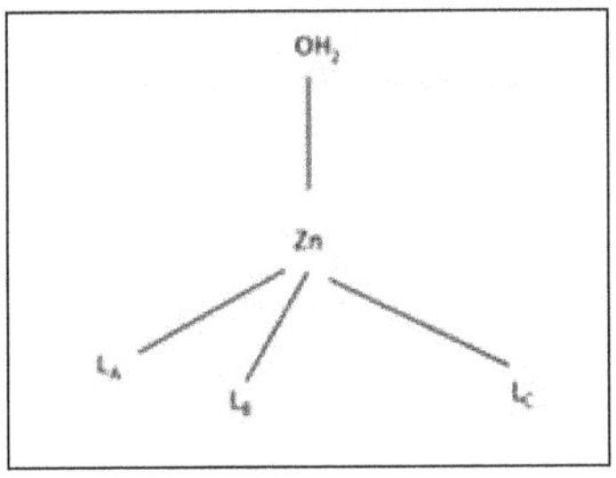

Coordination of zinc in a catalytic zinc site.

The first two ligands, LA and LB are separated by short spacers of 1 to 3 amino acids. L_C is separated from L_A and L_B by long spacers consisting of 5 to 196 amino acid residues, generally in the C- terminal end of the protein. Catalytic potentials of zinc enzymes are closely related to the nature of short and long amino acid spacers and the environment that they create for metal ligands.

Two common features are observed in all catalytic zinc sites. Firstly, the zinc ion is surrounded by a shell of hydrophilic groups that is embedded within a larger shell of hydrophobic groups. Secondly, the amino acid side chains coordinated to zinc ion often have hydrogen bonding interactions with other residues. These interactions have been proposed to orient the metal ligands, enhance the electrostatic interaction between metal and ligand and modulate the zinc water pK_a.

Zinc (II) is a borderline hard acid and does not have a strong preference for coordinating with either oxygen, nitrogen or sulfur atoms. Hence in enzymes, zinc (II) shows variable coordination, and is coordinated to the nitrogen of histidine amino acid residue, the oxygen of glutamate amino acid residue or aspartate amino acid residue and the sulfur of cysteine amino acid residue of the protein chain. Histidine is the most frequently observed ligand in a catalytic zinc site. The histidine ligands coordinate to zinc mostly through the N_ε atom and sometimes with N_δ atom. Carboxylate- zinc interactions with syn-stereochemistry are observed more frequently than those with antistereochemistry and the zinc ion prefers to be in the plane of the carboxyl.

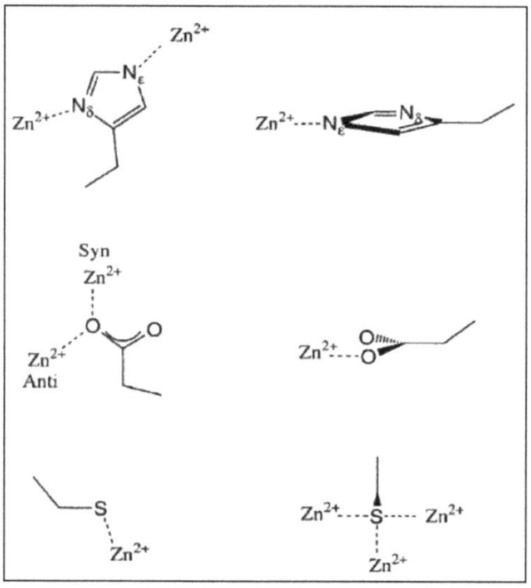

Mode of coordination of ligands to zinc ion.

The catalytic site of zinc metalloenzyme carbonic anhydrase II has zinc ligated to the imidazole nitrogen atoms of three histidine residues. His-94, His-96 are coordinated to zinc through their $N_\varepsilon 2$ atoms and His-119 is bound via $N_\delta 1$ atom. The fourth coordination site is occupied by a water molecule.

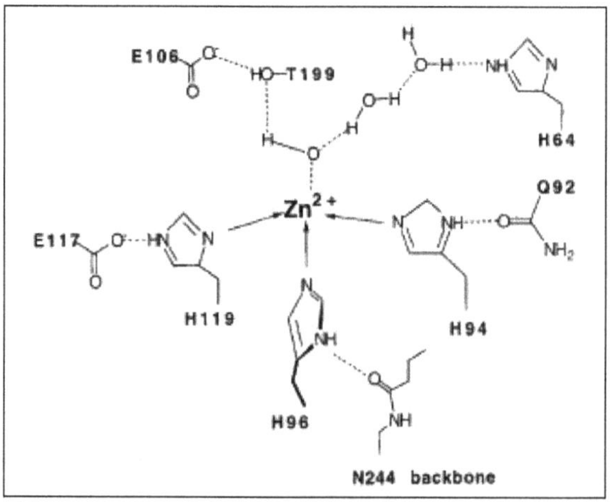

Active site structure of carbonic anhydrase II.

Role of Water Coordinated to Zinc

The coordinated water molecule is important for the catalytic action. The bound water can be either ionized to zinc coordinated hydroxide or polarized by a base from a nearby amino acid of enzyme or be replaced by the substrate. The preferred pathway for activating water is determined by the identity of the other three ligands and their spacing.

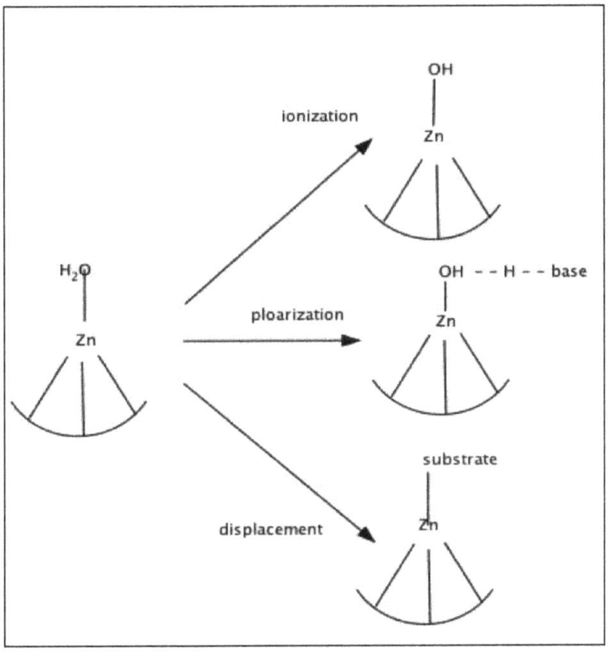

Role of zinc coordinated water in catalytic sites.

Mechanistic Pathways for Catalytic Action of Zinc Enzymes

Two general mechanistic pathways have been proposed for the catalytic action of zinc enzymes. In the zinc hydroxide mechanism, the hydroxide produced acts as a nucleophile that can attack the carbonyl carbon atom.

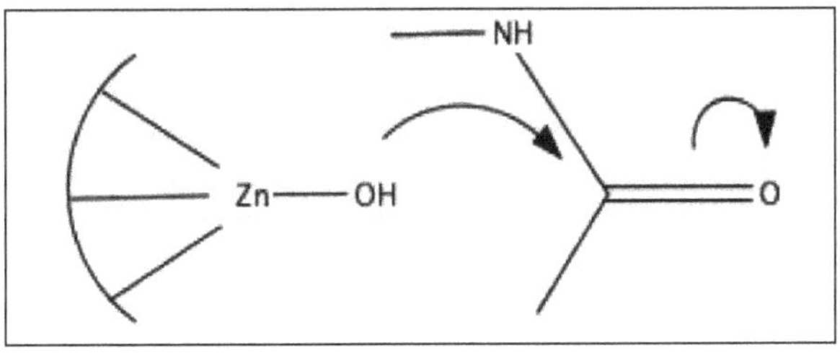

Zinc Hydroxide Mechanism.

In the zinc carbonyl mechanism, the substrate attaches to the zinc ion through a carbonyl group which is then polarized inducing nucleophilic attack at the carbon atom of the carbonyl group.

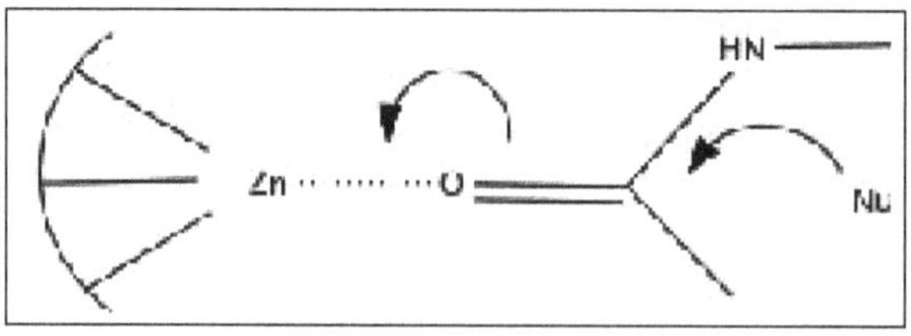

Zinc carbonyl mechanism.

The mechanism of action of each enzyme is decided by the amino acid ligands, their spacing in the protein sequence and the vicinal properties of the active centre created by protein folding.

Cocatalytic Zinc Sites

These occur only in multimetal zinc enzymes. Before the X-ray structure detection of these enzymes, the functions of these metal atoms were dubbed as "modulating" or "regulatory". In a cocatalytic zinc site there are two or more metal atoms, one, two or all of which can be zinc. They are in close proximity to one another and operate as a functional unit. One metal plays the catalytic role and the other metal atoms enhance the catalytic activity of the site.

Examples of enzymes that contain cocatalytic zinc sites are alkaline phosphatase (with zinc atoms and one magnesium atom), phospholipase (three zinc atoms), nuclease P1 (three zinc atoms).

In such zinc sites, a bridge is formed between the two participating metal atoms by a single amino acid residue like aspartate or glutamate which bind simultaneously to the two atoms. Aspartate and glutamate have two electronegative oxygen atoms attached to carbon hence they can easily serve as bridging ligands. In all the examples given above one zinc atom is catalytic while the remaining pair of metal atoms zinc and zinc or magnesium are linked by the bridging amino acid.

The cocatalytic metal site differs from the catalytic site in the kind of the ligands, number of ligands and hence geometry of the complex as well as the presence of a bridging ligand in cocatalytic site. The catalytic zinc site contains generally histidine; the cocatalytic zinc sites have aspartate and histidine. Some multizinc enzymes have serine, threonine, lysine and tryptophan as ligands in the cocatalytic zinc site. The common coordination number for zinc in catalytic site is 4 (geometry: distorted tetrahedral) whereas zinc in cocatalytic site has coordination number of 5 (geometry: slightly distorted trigonal bipyramidal). The spacing between the ligands in cocatalytic zinc site is irregular.

In phospholipase C, His^{-128}, His^{-142}, and Glu^{-146} are the ligands to Zn3, the catalytic zinc site. Aspartate bridges Zn1 and Zn2. Zn2 is bound to the bridging ligand Asp^{-122} and to His^{118}, His^{-69}, and Asp^{-55}. Zn1 is bound to the bridging ligand Asp^{-122} and to the amino and carbonyl groups of Trp^{-1}, the nitrogen of His^{-14}, and an oxygen of a water molecule which is an additional bridge for the two metal sites.

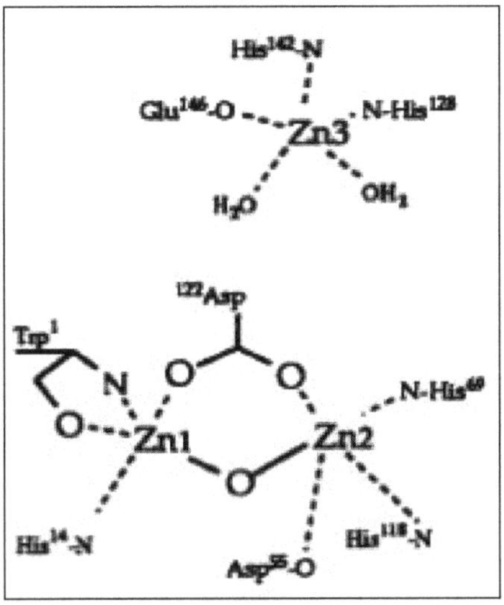

Phospholipase C.

Structural Zinc Sites

Zinc can play a structural role in certain proteins. In structural zinc sites, the zinc ion mainly stabilizes the tertiary structure. It has a symmetrical tetrahedral geometry satisfied by four amino acid side chains. There is no solvent coordinated to the metal. The amino acid which is mostly coordinated to the zinc atom in these sites is cysteine, with histidine also being present in many cases. The spacing between the coordinated ligands is irregular.

A zinc finger is a small protein structural motif in which coordinated zinc plays a structural role contributing to the stability of the motif. Zinc fingers are structurally diverse and are present among proteins that perform a broad range of functions in various cellular processes, such as replication and repair, transcription and translation, metabolism and signaling, cell proliferation and apoptosis.

Aspartate carbamoyltransferase is the only enzyme containing only a structural zinc atom. Alcohol dehydrogenase has two zinc atoms, one a catalytic and other a structural zinc atom. Four cysteins coordinate the structural zinc in a tetrahedral geometry.

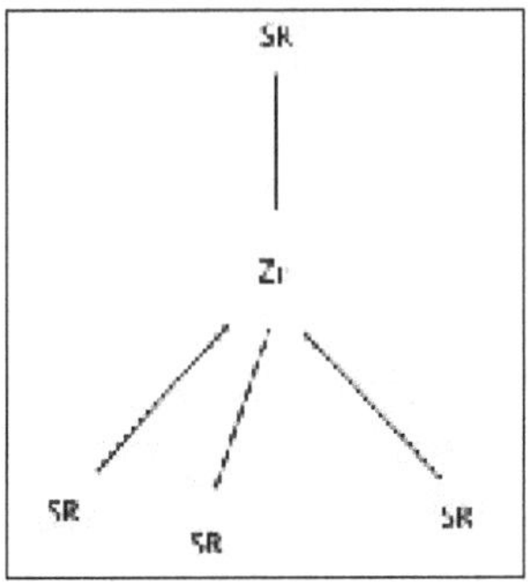
Structural zinc site in alcohol dehydrogenase.

Zinc Enzymes

Over 200 zinc enzymes are known (NRC, 2003). Zinc activates enzymes of protein metabolism, is part of carbonate dehydratase and carbon dioxide transfer, and is important for immune function.

Carboxypeptidase A (CPDA)

Carboxypeptidase A is one of the most studied peptide hydrolyzing enzyme. In 1929 Waldschmidt – Leitz and Purr isolated CPDA for the first time. It was the first metalloprotease and second zinc enzyme to be identified.

There are two kinds of carboxypeptidases. Enzymes of one class are not metalloenzymes. Yeast carboxypeptidase C belongs to this class. The second class of carboxypeptidases are metalloenzymes. Carboxypeptidase A (CPDA) is an example of this class of enzyme. It is an exopeptidase (i.e. one which cleaves only terminal peptide bonds) that hydrolyses a peptide bond at carboxyl terminal (C- terminal) end of a peptide or protein. A related protein carboxypeptidase B catalyzes the hydrolysis of basic amino acids lysine, arginine, ornithine from C-terminal position of polypeptides. Hydrolysis by CPDA occurs only if the C-terminal residue has S- configuration and its carboxyl group is free. Peptides possessing any C- terminal residue except proline will be hydrolysed and an aromatic amino acid side chain at the C- terminal residue is favoured in the substrate.

There are three forms of bovine pancreatic carboxypeptidase A; A_α, A_β and A_γ ; with 307, 305 and 300 amino acid residues respectively. The A_β and A_γ are shortened at the N- terminus. The reaction for hydrolysis of peptides by CPDA is:

$$R-CO-NH-CHR'-COO^- + H_2O \rightleftharpoons R-COO^- + H_3N^+ -CHR'-COO^-$$

At neutral pH, in the absence of enzyme, the above reaction is slow with rate constants of the order of 10^{-11} s^{-1}. However, if the above reaction is catalyzed by carboxypeptidase the rate constants can attain values of 10^4 s^{-1}. CPDA is initially produced in an inactive form (zymogen), procarboxypeptidase A, in the pancreatic juice of animals. The inactive form is converted to the active CPDA by the enzyme enteropeptidase.

Active Site of Carboxypeptidase A

CPDA consists of a single protein chain (M= 34,300) and has one Zn^{2+} ion present in the active site. It is ellipsoidal with a cleft on one side that contains the zinc ion. In the native enzyme, the zinc coordination number is five (imidazole nitrogens from two histidine residues, histidine- 69 and histidine-196, two carboxylate oxygens of glutamate–72 and a water molecule) with a distorted pentacoordination geometry. Many acidic and basic amino acids namely asparagine-144, arginine-145, tyrosine-248, arginine-127 and glutamate-270 are also present in the active site and are important for the activity of the enzyme. The active site cavity has a hydrophobic pocket that accommodates the non-polar R' of the C- terminal amino acid of the substrate undergoing hydrolysis.

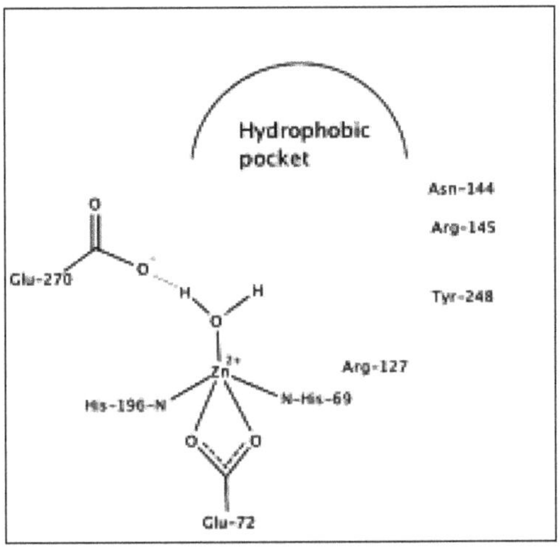

Active site of Carboxypeptidase A.

Catalytic Mechanism of Carboxypeptidase A

The mechanism for hydrolysis of peptides or proteins by carboxypeptidase A was proposed by Christianson and Lipscomb. In this mechanism, the water molecule coordinated to zinc is important for the catalytic action. The bound water is converted to nucleophilic hydroxide, by interaction with glutamate-270, which then attacks the carbonyl carbon of the scissile peptide bond.

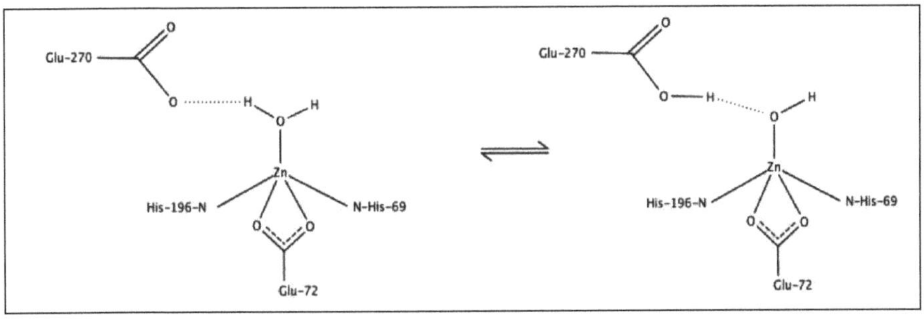

Interaction of coordinated water with glutamate-270 present in the active site of CPDA.

The first step is positioning of the substrate on the enzyme Figure. The enzyme interacts with the substrate via hydrogen bonds (arginine-145 and asparagine-144 with terminal carboxylate group, tyrosine-248 with the penultimate peptide NH, arginine-127 with carbonyl oxygen of the terminal peptide bond) and van der Waals attraction if the terminal group is aromatic (hydrophobic pocket with hydrophobic R' of C- terminal amino acid). The interaction of arginine-127 with the carbonyl of the terminal peptide bond increases the Lewis acidity of the latter making it more susceptible for nucleophilic addition. Then the water molecule bound to zinc performs a nucleophilic attack on the the carbonyl carbon of the C-terminal peptide bond with the simultaneous transfer of a proton to glutamate-270.

At this point the glutamate-72 bound to zinc ion becomes monodentate from bidentate, maintaining the five coordination of zinc ion. This process is called carboxylate shift. The tetrahedral intermediate formed by the nucleophilic attack is shown in Figure. The protonated glutamate-270 now transfers its proton to the amide group of the terminal peptide bond. This causes the collapse of the tetrahedral intermediate resulting in the formation of the free carboxyl terminal amino acid and the shortened peptide chain with its terminal carboxylate group bound to zinc Figure. A water molecule adds to the zinc ion. Finally the cleaved peptide leaves and zinc ion gets back to its original pentacoordination.

Mechanism of hydrolysis of a peptide by Carboxypeptidase A.

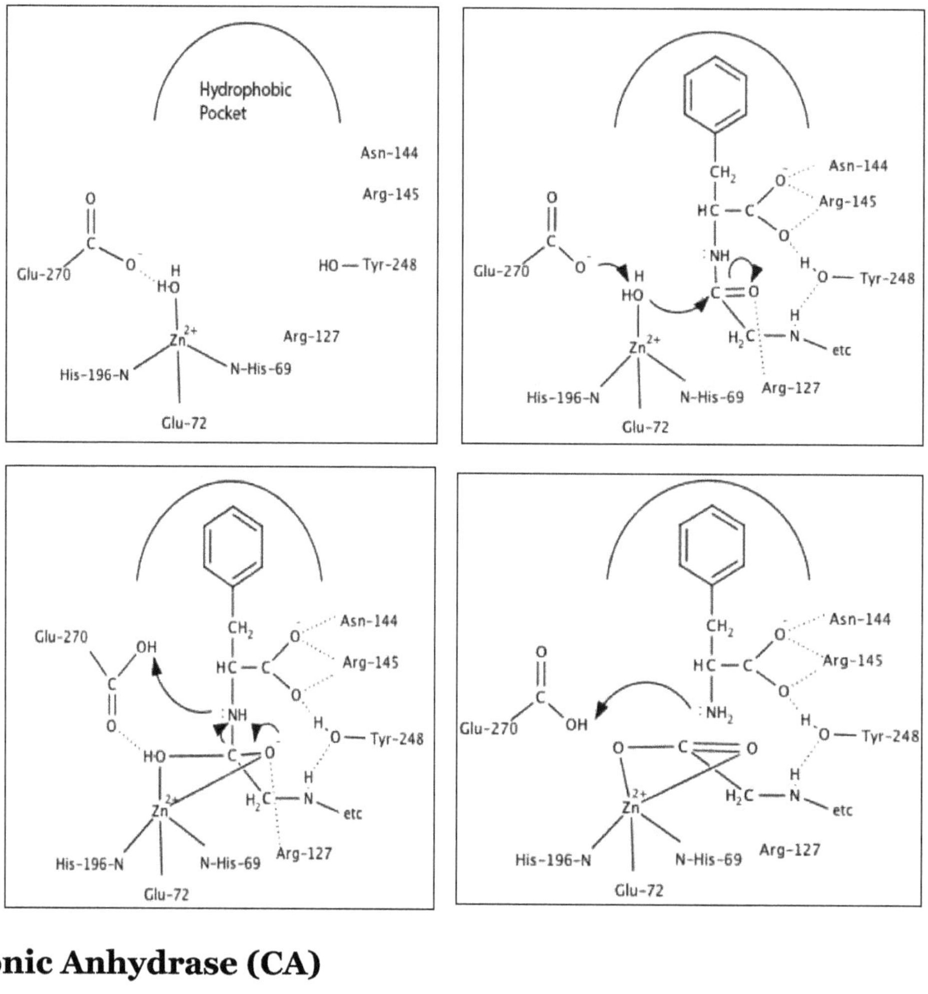

Carbonic Anhydrase (CA)

It is a lyase that catalyzes the reversible hydration of carbon dioxide to form the bicarbonate ion:

$$CO_2 + H_2O \rightleftharpoons HCO_3^- + H^+$$

In the absence of the enzyme this reaction occurs relatively slowly at physiological pH and the presence of CA increases the rate upto a million fold. In vitro it catalyzes several other reactions also such as the hydrolysis of esters and the hydration of aldehydes.

Carbonic anhydrases are found in all kingdoms of life, occurring in animals, plants and several bacteria. They play an essential role in biological processes like respiration, photosynthesis and in maintaining pH. There are several forms of CA. All the forms are single chain polypeptides with molar mass about 30,000 Da and have one zinc ion per molecule. High-activity forms have k_{cat}~10^6 s^{-1} at 25° C and are labeled II. Low- activity forms and the very-low-activity forms have k^{cat}~10^5 s^{-1} and 10^3 s^{-1} and are labeled as I and III respectively. The best studied carbonic anhydrase is CA II from red blood cells.

Active Site of Human Carbonic Anhydrase II (HCA II)

The active site of human carbonic anhydrase II is located at the bottom of a cleft in the enzyme. The protein is composed of 259 amino acids. HCA II contains an important cofactor at the active site, namely, dipositive zinc ion, that helps activate water molecules prior to their reaction with CO_2. This zinc is coordinated to the imidazole nitrogen atoms of three histidine residues. Histidine94, histidine-96 are ligated to zinc via their Nε atoms and histidine-119 is bound via Nδ atom. The three histidine NH protons are all engaged in hydrogen bonding. Histidine-94, histidine-96, histidine-119 are involved in hydrogen bonding with glutamine-92, asparagine-244 and glutamate117 respectively. The fourth coordination site of zinc is occupied by a water molecule to give a distorted tetrahedral geometry. The active site pocket also contains ordered network of water molecules and other amino acids like threonine-199, glutamate-106, histidine-64 that are important for hydrogen bonding, proton transfer (rate determining step in catalytic mechanism) and for binding the CO_2 substrate.

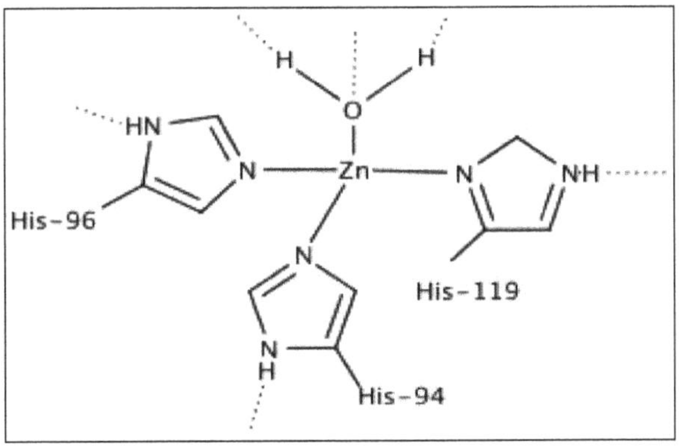

Active site of Human Carbonic Anhydrase (II).

Catalytic Mechanism of Human Carbonic Anhydrase II (HCA II)

Zinc(II) is a Lewis acid. Zinc ion present in HCA(II) is more acidic than the zinc in CPDA. The difference is because of the ligands coordinated to zinc. CPDA has two histidines and a bidentate glutamate ion ligated to zinc. HCA(II) has three histidines. Histidine being neutral and less basic than glutamate anion, makes the zinc more acidic in HCA(II). The binding of water to zinc, reduces the pKa of water from its normal 14 down to 7. This leads to its deprotonation at neutral pH to form the strong hydroxide nucleophile.

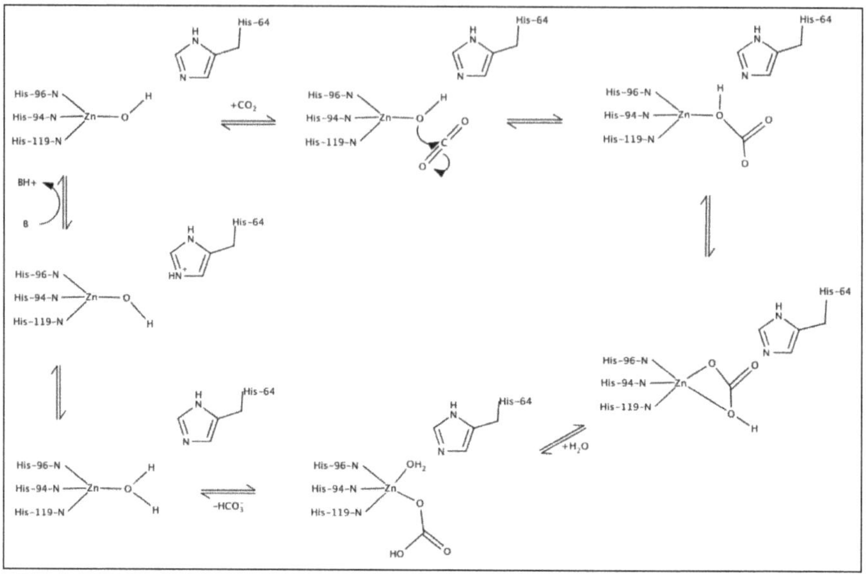

Catalytic mechanism of human carbonic anhydrase (II).

The dissociated proton is transferred indirectly to histidine-64 via water molecules. Histidine-64 shuttles protons from the active site to the buffer in solution, shifting the equilibrium in the forward direction in favor of the formation of hydroxide ion. The enzyme then positions carbon dioxide for nucleophilic attack by the hydroxide. The nucleophilic attack by hydroxide on carbon dioxide produces a zinc bound bicarbonate. Addition of water displaces the bicarbonate ion, completing the cycle.

Alkaline Phosphatase (ALP)

Origin and Distribution of Alkaline Phosphatase (ALP)

The homodimeric metalloenzyme known as alkaline phosphatase [phosphate-monoester phosphohydrolase (alkaline optimum); EC 3.1.3.1] is a hydrolase enzyme with 449 amino acid residues. It removes phosphate groups from many types of organic phosphates (nucleotides, proteins) to provide phosphate required for bone growth. Alkaline phosphatase is found in all kingdoms of life (plants, animals and bacteria) with the exception of some plants.

It is the first zinc enzyme to be discovered which has a catalytic zinc centre that contains more than one kind of metal atom. It has a mass of 94 kilodalton. Each monomer of ALP binds metal ions. The metal ions they carry can differ, although zinc and magnesium are the most common. ALP catalyzes hydrolysis of a variety of phosphate monoesters.

$$R-OPO_3^{2-} + H2O \rightleftharpoons ROH + HOPO_3^{2-}$$

It also catalyzes transphosphorylation reactions. In the presence of high concentrations of phosphate acceptors, enzyme ALP transfers phosphate to alcohol to form another phosphate monoester.

Alkaline Phosphatase is membrane bound and found in liver, bone, intestine, kidney and placenta. The liver, bone and kidney enzymes are different forms of the same gene product, the intestine and placental forms are isoenzymes. Physiologically increased levels of ALP are seen during periods of active bone growth, particularly in infants and at puberty. The optimum pH for the activity of ALP is in the mild alkali region.

Protein Structure of Alkaline Phosphatase (ALP)

The overall dimension of ALP is approximately 100 Å * 50 Å * 50 Å with the active sites 30 Å apart from each other on opposite sides of the molecule. Each active site of the dimeric enzyme contains three metal binding sites, two are occupied by zinc(II) ions and one by magnesium(II) ion. The two zinc ions are 4 Å apart and the magnesium ion is approximately 7 Å from Zn1 and 5 Å from Zn_2 in each of the subunits. The two subunits in the dimer are held together by hydrogen bonds, hydrophobic interactions and salt bridge.

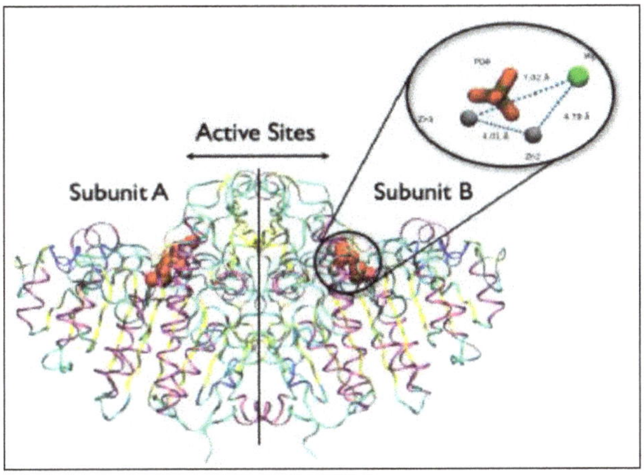

Crystal structure of Alkaline Phosphate in ribbon representation, coloured by secondary structure elements, with a close-up showing inorganic phosphate and metal ions in licorice representation and coloured by the atom type.

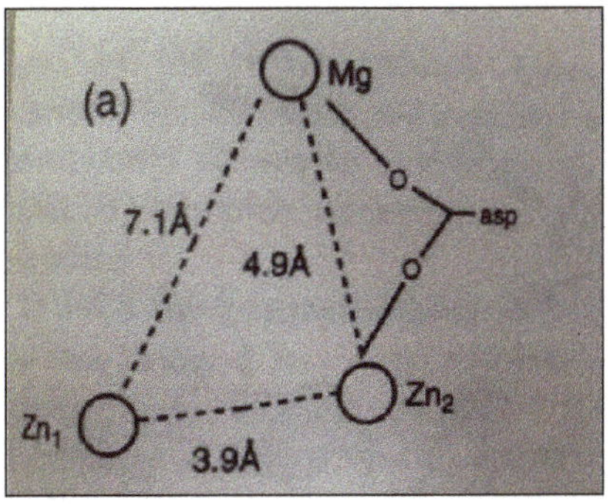

Metal Ions involved in the active site of each subunit of homodimeric enzyme ALP.

Comparison of the amino acid sequences from a number of ALPs shows that the enzymes are very well conserved, especially near the active site. The residues Ser-102 and Arg-166 are conserved in all the cases. These amino acid residues interact directly with the phosphate ester. The X-ray crystal structure of ALP complexed with inorganic phosphate was determined. The first zinc atom (Zn_1) has a coordination number of five with tetragonal pyramidal geometry. Zn_1 binding site consists of His-331, His-412 and an Asp-327 residue. Both carboxyl oxygens of Asp-327 are coordinated to Zn_1 along with imidazole nitrogen atoms of histidines. Zn_1 is also coordinated to one of the oxygens of the phosphoester. The second zinc (Zn_2) has a distorted tetrahedral geometry. Zn_2 binding site consists of Asp-369, Asp-51 and His370 and one oxygen of phosphoester. One of the carboxyl oxygens of Asp-369 and Asp-51 coordinate Zn_2. The nitrogen atom of His-370 is also coordinated to Zn_2. The magnesium (II) ion has a slightly distorted octahedral coordination environment. Second carboxyl oxygen of Asp-51 is coordinated to Mg(II). Mg(II) is also coordinated to two oxygen atoms from Glu-322, Thr-155 and three oxygen atoms from water molecules.

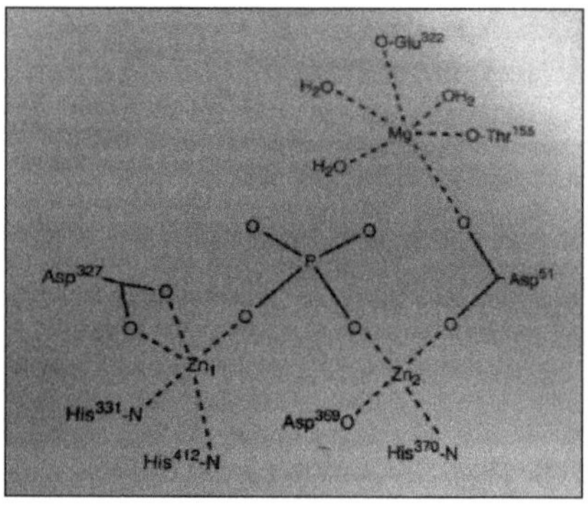

Structure of ALP complexed with inorganic phosphate.

Carboxylates can act as terminal (η^1), as chelating (η^2) or as bridging μ-η^1: η^2 ligands. All these three types of carboxylate to metal coordination are found in Alkaline Phosphatase. η^1 is shown by Aspartate-369, η^2 is exhibited by aspartate-327 and the bridging mode is shown by Asp-51.

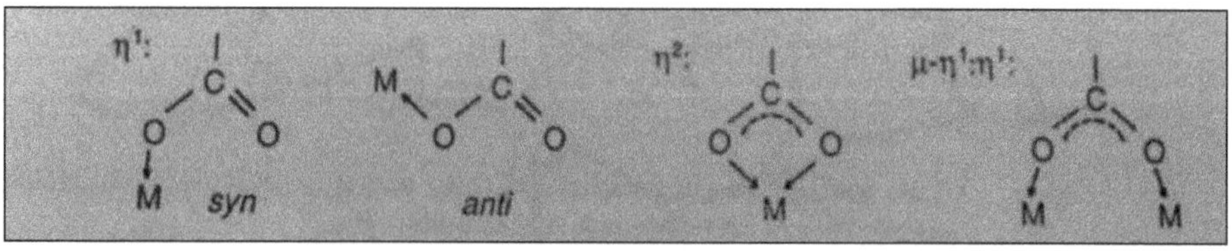

Modes of binding of carboxylates to metal.

Mechanism of Alkaline Phosphatase (ALP)

Two Two-Metal Ion Assisted Catalytic Mechanism

In 1991, Kim and Wyckoff proposed a two two-metal ion assisted catalytic mechanism of phosphate monoester hydrolysis. The proposed reaction mechanism, involves two zinc metal ions, two amino acid residues of the enzyme, Arg-166 and Ser-102 and one out of three water molecules present in the active site. They arrived at the mechanism based on the structure of ALP with bound phosphate. ALP functions through formation of a phosphoseryl intermediate to produce free inorganic phosphate or by transferring the phosphoryl group to other alcohols.

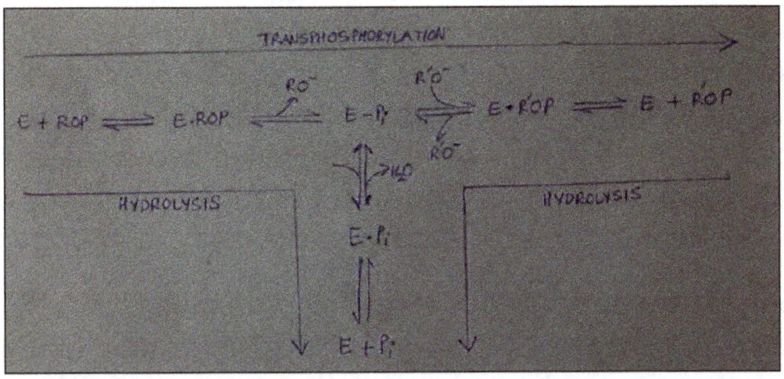

Overall reaction scheme of the reaction where both hydrolysis and transphosphorylation reactions catalyzed by ALP are shown.

The free enzyme interacts with the dianion of the phosphomonoester (ROP) to form the enzyme substrate complex (E.ROP). In this complex the ester O is coordinated to Zn_1; a second phosphate O is coordinated to Zn_2 and the remaining phosphate oxygens interact with the guanidinium group of Arg-66 by hydrogen bonds. Then Ser-102 activated by Zn_2 acts as a nucleophile and attacks the phosphate group of the substrate forming the covalent phosphoenzyme intermediate (E-Pi) and liberating alkoxide. Now a water molecule coordinates to Zn_1 in the position vacated by RO-. In alkaline pH this water gives Zn-OH. This activated water hydrolyzes the covalent intermediate (E-Pi) into noncovalent enzyme-phosphate complex (E.Pi). The dissociation of phosphate from this complex gives the free enzyme. This is the slowest step and therefore rate determining. The Mg(II) ion does not have a direct role in catalysis but it is important for full enzyme activity.

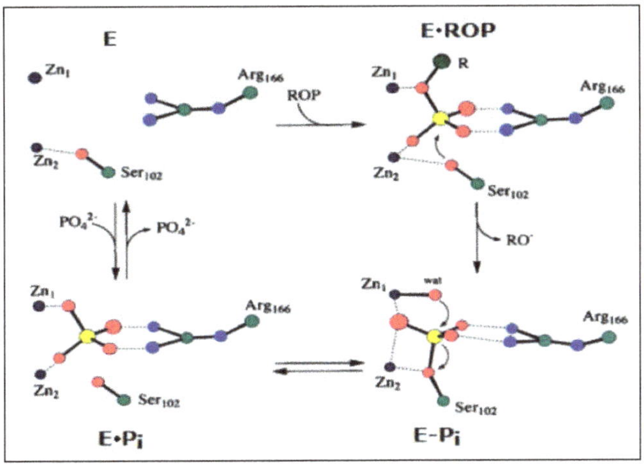

Major intermediates in the proposed mechanism of ALP (Two Metal ion catalysis).

Three-Metal Ion Assisted Catalytic Mechanism

In 2000, a revised catalytic mechanism (Three metal ion catalysis) of hydrolysis of phosphate monoester by ALP. It could explain two uncleared aspects of the Two Metal ion catalysis mechanism. These include the mechanism of deprotonation of the hydroxyl group of Serine-102 and the role of magnesium ion in the third metal site. Three water molecules are present in the phosphate binding site in the free enzyme (E).Magnesium has a coordination number of six and is coordinated to oxygen atoms of Asp-51, Thr-155 and Glu-322 and three oxygen atoms from the three water molecules different from the earlier ones. One of the coordinated water molecules is at a distance of 4.7 Å from Zn2 and 3.1 Å from Ser-102. This water is coordinated as a hydroxide ion to magnesium and functions as a general base and accepts the proton from Oγ of Ser-102.

In the E.ROP complex the ester oxygen atom is coordinated to Zn1. Two oxygen atoms of the ester substrate are hydrogen bonded to the guanidinium group of Arg-166 and the fourth oxygen atom is coordinated to Zn_2. Ser-102 occupies the position opposite to the leaving group. The hydroxide ion coordinated to magnesium deprotonates Ser-102 by acting as the general base. Then Ser-102 performs a nucleophilic attck on the phosphorus atom resulting in the formation of the covalent enzyme-phosphate intermediate (E-P). This results in the inversion at the phosphorus centre and the loss of the leaving group (RO-).

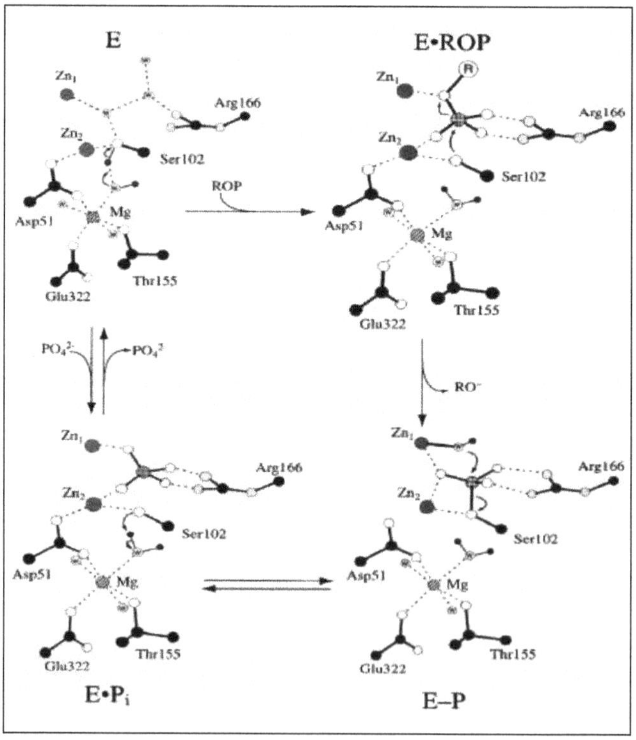

Three-metal ion assisted catalytic mechanism of Alkaline Phosphatase.

This is followed by the attack on the phosphorus in E-P by the nucleophilic hydroxide ion coordinated to Zn_1, resulting in the formation of the non-covalent enzyme-phosphate complex (E.Pi). A second inversion of configuration occurs at the phosphorus centre. The water molecule coordinated to magnesium now acts as a general acid, donating a proton to Ser-102.

Liver Alcohol Dehydrogenase (LADH)

Alcohol Dehydrogenases belong to the oxidoreductases family of enzymes. The transfer of hydride is a redox reaction but it can also be visualized as nucleophilic attack of hydride on the substrate. If the process is carried out enzymatically the enzyme provides the binding sites for the two moieties between which transfer of hydride is occurring. Liver Alcohol Dehydrogenases (LADH) are one such metallo-enzymes. A great deal of structural and mechanistic studies have been performed on the horse liver alcohol dehydrogenases. LADHs catalyze the reversible dehydrogenation/ oxidation of primary and secondary alcohols to aldehydes and ketones respectively.

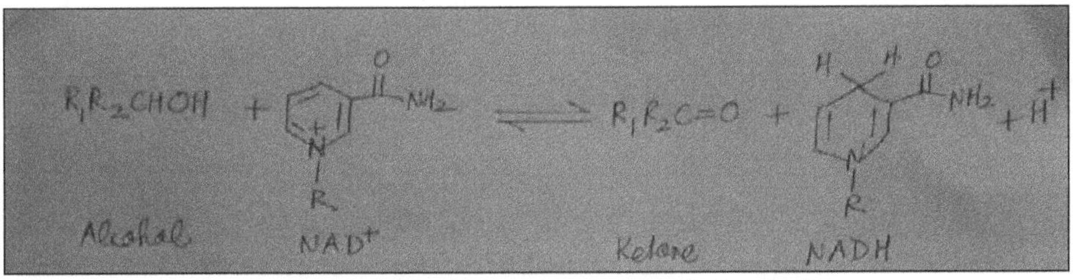

Reaction catalyzed by the enzyme Liver Alcohol Dehydrogenase.

Structure of Liver Alcohol Dehydrogenase

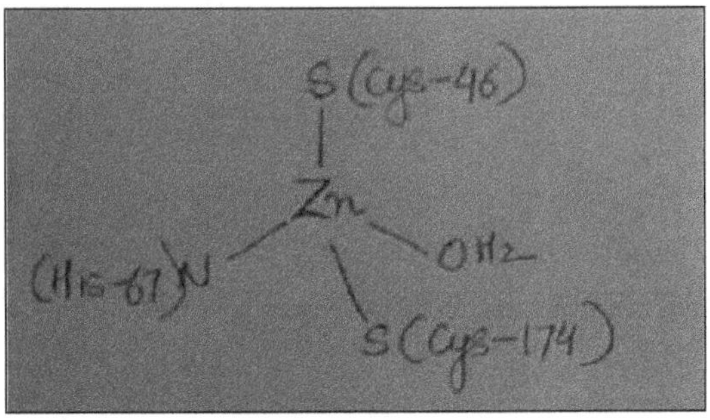

Catalytic Zinc in LADH.

LADH consists of two identical polypeptide subunits each of mass 40 kilodalton. Each subunit contains two zinc(II) ions and binds one mole of NAD+. Each subunit of the dimeric enzyme is divided into a coenzyme binding domain and a catalytic domain that are separated by a cleft containing a deep pocket. Both zinc ions are present in the catalytic domain, one is a catalytic zinc and the second zinc has a structural role. The catalytic zinc(II) is located at the bottom of a hydrophobic pocket in a deep cleft about 2 nm from the surface of the protein. It has a coordination number of four and is ligated to two cysteinate sulphur atoms from Cys-46 and Cys174, one Nε2 nitrogen of His-67 and a water molecule. The coordination geometry is distorted tetrahedral.

The structural zinc is close to the protein surface, but inaccessible to solvent. It is coordinated to four cysteine thiolate ligands (Cys-97, Cys-100, Cys-103, Cys-111) in a regular tetrahedral arrangement.

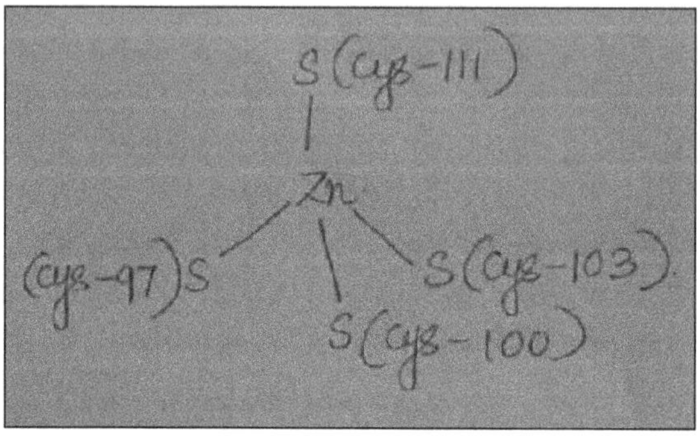

Structural zinc in LADH.

Catalytic Mechanism of Liver Alcohol Dehydrogenase

Each monomer has two protein domains (catalytic domain and coenzyme binding domain) which can rotate freely with respect to each other. Solvent is accessible to the catalytic zinc and the fourth ligand is water in the open conformation. When the coenzyme NADH binds it triggers a major change in the conformation of the enzyme. The binding of NADH to the enzyme induces rotation of one domain resulting in the closed form. This leads to the proper orientation of the groups. The catalytic zinc ion can now bind the substrate (aldehyde or ketone) in such a way that the CH_2 group of nicotinamide ring of NADH faces the carbonyl carbon. Thus catalytic zinc(II) has two roles, geometric and electronic. It gives a specific spatial orientation to the substrate and polarizes the carbonyl bond.

The catalytic mechanism for Liver Alcohol Dehydrogenase is shown in Figure below. The coenzyme NAD+ binds to the coenzyme binding domain. At this stage the enzyme is in its open form. The substrate alcohol enters the cleft between the catalytic domain and the coenzyme binding domain and displaces the water molecule coordinated to the catalytic zinc. The rotation of domain changes the LADH enzyme into the closed form. All the water molecules are extruded out from the active site. Then the oxygen bound proton is removed from the zinc coordinated alcohol. This is followed by transfer of hydride from the alcohol CH to the nicotinamide ring. The intermediate involved is a ternary complex of NADH-Aldehyde-Catalytic zinc. The product aldehyde leaves and the catalytic zinc(II) retains its original coordination by attaching a water molecule. NADH is released and this brings back LADH to its open form.

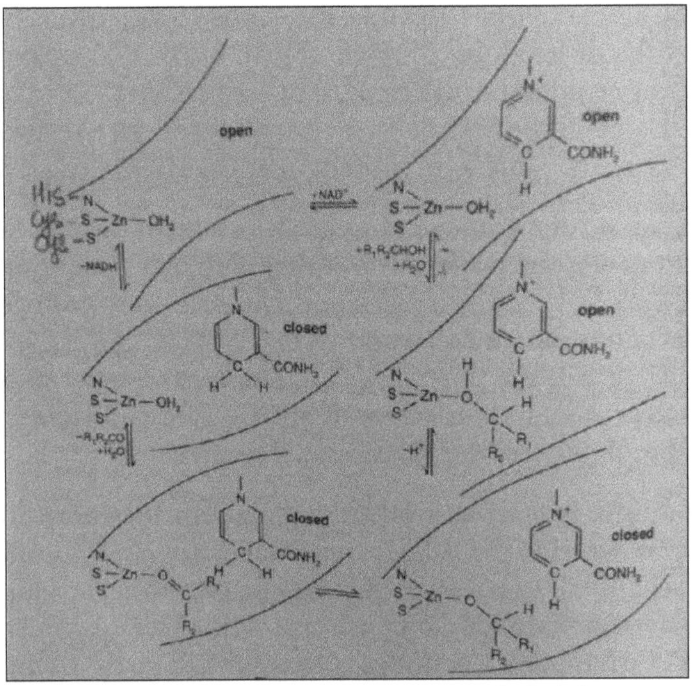

Catalytic Mechanism of Dehydrogenation of alcohols by LADH.

Zinc Finger Proteins

Zinc finger (ZnF) proteins are a massive, diverse family of proteins that serve a wide variety of biological functions. Due to their diversity, it is difficult to come up with a simple definition of what unites all ZnF proteins; however, the most common approach is to define them as all small, functional domains that require coordination by at least one zinc ion. The zinc ion serves to stabilize the integration of the protein itself, and is generally not involved in binding targets. The "finger" refers to the secondary structures (α-helix and β-sheet) that are held together by the Zn ion. Zinc finger containing domains typically serve as interactors, binding DNA, RNA, proteins or small molecules.

ZnF Protein Families

Cys_2His_2 was the first domain discovered (also known as Krüppel-type). It was initially discovered as a repeating domain in the IIIA transcription factor in Xenopus laevis. IIIA has nine repeats of the 30 amino acids that make up the Cys_2His_2 domain. Each domain forms a left-handed ββα secondary structure, and coordinates a Zn ion between two cysteines on the β-sheet hairpin and two histidines in the α-helix, hence the name Cys_2His_2. These resides are highly conserved, as well as a general hydrophobic core that allows the helix to form. The other residues can show great sequence diversity. Cys_2His_2 zinc fingers that bind DNA tend to have 2-4 tandem domains as part of a larger protein.

The residues of the alpha helices form specific contacts with a specific DNA sequence motif by "reading" the nucleotides in major groove of DNA. Cys_2His_2 proteins are the biggest group of transcription factors in most species. Non-DNA binding proteins can have much more flexible tertiary structure. Examples of Cys_2His_2 proteins include the Inhibitor of Apoptosis (IAP) family of proteins and the CTFC transcription factor.

Treble clef fingers are a very diverse group of ZnF protiens both in terms of structure and function. What makes them a family is a shared fold at their core that looks a little like a musical treble clef, especially if you squint. Most treble clef finger motifs have a β hairpin, a variable loop region, a β hairpin, and an α helix. The "knuckle" of the β hairpin and the α helix contain the Cys-x-x-Cys sequence necessary to coordinate the Zn ion. Treble clef fingers often form the core of protein structures, for example the L24E and S14 ribosomal proteins and the RING finger family.

Zinc ribbons are a little less structurally complex than the other two major groups. Zinc ribbons contain two zinc knuckles, often β hairpins, coordinating a zinc ion via a two Cys residues separated by 2-4 other residues on one knuckle, and a Cys-x-x-Cys on the other. Examples of zinc ribbon-containing proteins include the basal transcription factors TFIIS and TFIIB that for a complex with RNAPII to bind DNA, and the Npl4 nuclear core protein that uses a zinc ribbon to bind ubiquitin. Cys_2His_2, treble clef fingers, and zinc ribbons form the majority of zinc fingers, but there are several other smaller groups that don't fit neatly into these three.

Practical Uses for Zinc Finger Proteins

As soon as the specificity of ZnF proteins was understood, the idea of creating synthetic ZnF proteins became the focus of many biotech companies. Cys_2His_2 motifs each recognize a specific nucleotide triplet depending on the residues on their α helix. This was thought to form a simple code that could be used to recognize very specific DNA sequences by engineering specific ZnF motifs in tandem within a protein. Another domain of the protein could then serve some desired biological function once the ZnF bound the target sequence. For example, cutting at one specific point in the genome and inserting a transgenic element. But alas, it was not so simple. ZnF recognition residues also have cross-recognition to adjacent elements, thus each motif must be chosen in the context of those around it.

Xenopus Transcription Factor IIIA (TFIIIA)

Zinc finger protein was first recognized 30 years ago in Xenopus transcription factor II-IA(TFIIIA), which activates the transcription of the 5s RNA gene, in the African clawed toad Xenopus.

Structure of Zinc Finger in TFIIIA

The primary structure of TFIIIA contains nine repeat sequences of about 30 amino acids and each repeat contains two cysteine residues, two histidine residues and three hydrophobic residues, in conserved positions. The two cysteine and two histidine residues per 30 amino acid unit form a tetrahedral coordination complex with zinc atom, generating peptide domains. These peptide domains have a typical, compact folding. The DNA-binding loop formed by the coordination of these ligands by zinc has a protruding finger like shape, thus the name zinc finger. TFIIIA has nine zinc fingers.

The characteristic sequence for each domain is -(Tyr,Phe)-X-Cys-$X_{2\text{-}4}$-Cys-X_3-Phe-X_5-Leu-X -His-$X_{3\text{-}5}$-His where X is a variable amino acid.

The domains have a characteristic β-sheet structure on the cysteine side and a α-helical arrangement on the histidine side. The spacing of three to four amino acids between the two coordinating histidine residues is important because it places these amino acids on the same face of the helix in an ideal position for coordination to zinc. The coordination of two cysteine residues (which are two to four amino acids apart) with zinc induces the formation of N-terminal antiparallel ββ loop structure in TFIIIA.

This proposal of structure of peptide domain in TFIIIA was supported by experimental data. The cysteine histidine coordination sphere for Zn in TFIIIA was confirmed by EXAFS spectroscopy. Single zinc finger peptides have been prepared and purified. Such units have been found to bind one equivalent of Zn^{2+} or Co^{2+}. Fe^{2+} and Ni^{2+} also bind to the peptide. Metal ion titration experiments have shown that Zn^{2+} binds 10^3 to 10^5 times more strongly than other ions. Since zinc is spectroscopically inactive, Uv-Vis spectroscopic studies on the cobalt complex have been done and have showed tetrahedral coordination for Co. CD spectra of the peptide without Zn or Co showed the absence of a well-defined structure. Addition of either one of the metals leads to a change in the CD spectra.

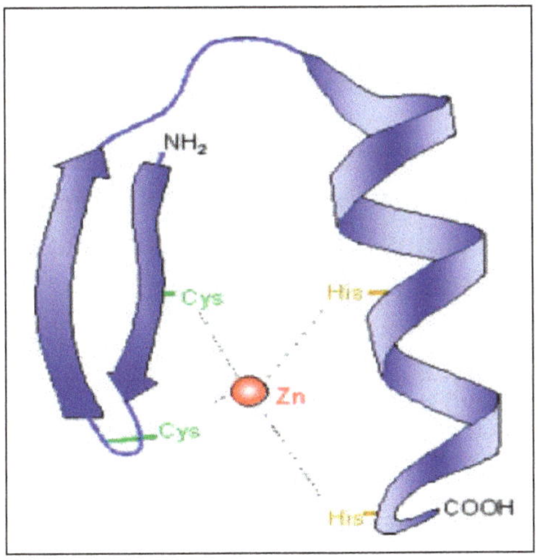

Structure of one zinc finger in TFIIIA.

Role of Zinc in Zinc Finger Protein TFIIIA

Zinc is crucial for the functioning of these transcription factors. It was observed that loss of zinc in TFIIIA leads to a significant change in protein conformation as well as loss of specific DNA binding ability. This led to the conclusion that the role of Zn in TFIIIA was to hold the protein in the proper conformation, so that specific amino acids side chains make specific contacts on the DNA binding site. Zinc brings the α-helix and β sheet closer together forming a hydrophobic core which is important for finger stability. The stability of the ββα domain is because of the intrastrand crosslinking provided by coordination with zinc.

The structures of very small protein domains are generally stabilized by the formation of disulphide bonds, or by binding to metal ions (most frequently, zinc). The binding of metal increases the thermal and conformational stability of small domains but is not directly involved in their function. Disulphide bridges can be easily cleaved by reduction in the reducing atmosphere of the cell and the nucleus, hence nature has chosen zinc rather than disulphide bridges for stabilization of structure of zinc fingers. Zinc is an ideal cation for structural role in proteins because of the following reasons:

- Natural abundance of zinc is high.
- It is completely non-toxic in nature.
- Zinc(II) is high in the Irving –Williams series and thus forms stable complexes especially with S and N donors.
- The zinc(II) ion has a filled d orbital and hence does not have any redox activity associated with it, that could promote DNA damage.
- Zn(II) can adopt a tetrahedral coordination. The tetrahedral coordination for zinc in zinc finger stabilizes it as higher coordination numbers will require more polypeptide organization which might present an entropic barrier to protein folding.

Classification of Zinc Finger Proteins

The rapid growth in the number of zinc finger proteins necessitates their classification. They can be classified in different ways. Of the DNA (or RNA)-binding variety, three major classes of zinc finger proteins have been established to date in eukaryotes, based on their unique and highly conserved consensus amino acid sequences. They are summarized in Table.

Zinc Finger Class	Consensus amino acid sequences	Examples
I (C_2H_2)	Cys-X_{2-4}-Cys-X_{12}-His-X_{3-5}-His	Xenopus TFIIIA
II (C_4)	Cys-X_{2-}Cys-Xn-Cys-X_2-Cys-Xn-Cys-X_2-Cys-Xn-CysX_2-Cys	Glucocorticoid receptor
III (C_6)	Cys-X_2-Cys-X_6-Cys-X_{5-12}-Cys-X_2-Cys-X_{6-8}-Cys	Candida albicans

Class I Zinc Finger (C_2H_2)

This class is also called the classical zinc finger. It comprises hundreds of zinc finger proteins that are found in eukaryotes ranging from yeast to humans. It is described as Cys-X_{2-4}-Cys-X_{12}-His-X_{3-5}-His where X is a variable amino acid. The two cysteine residues and two histidine residues are tetrahedrally coordinated to zinc as shown in Figure below. It has two to three β strands in its N-terminal sequence and one α-helix in the C-terminal half of X12H. C_2H_2 zinc finger proteins are nearly 1% of total mammalian proteins. These proteins bind DNA and some can also bind to RNA or protein. These proteins contain 2 or more repeating zinc finger units. GAGA factor is the only protein that contains a single C_2H_2 finger. Xenopus TFIIIA, human transcription factor Sp1, Mouse Zif268 are some examples of this class.

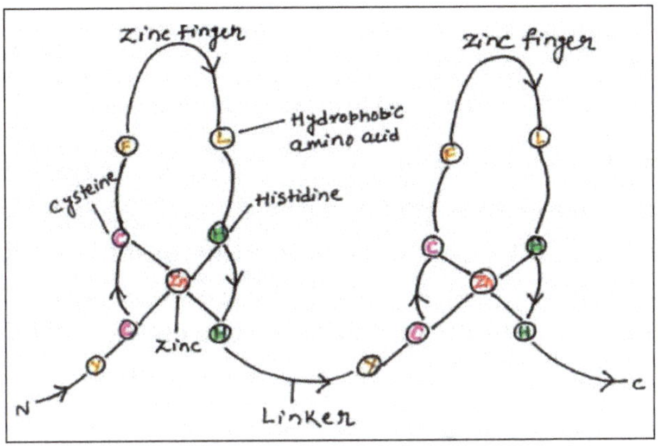

C_2H_2 zinc finger motif.

Class II Zinc Finger (C_4)

This class encompasses the Cys_4 (C_4) zinc fingers. The consensus amino acid sequence for this class is given in Table. The glucocorticoid (GR) and oestrogen (OR) receptors belong to this class of zinc finger proteins. These steroid receptors have DNA binding domain which contains 70 amino acids residues with eight conserved cysteine residues. DNA binding domain of both these receptors show two zinc atoms each coordinated to four cysteines. These zinc-binding sites fold to form a single structural domain. The DNA-binding site is located between the two zinc atom complexes. The DNA-binding recognition site derives from the α-helix that is part of the linking peptide region between the two zinc sites. Only one of the zinc based units plays a significant role in DNA recognition, the other helps in dimerization of the protein.

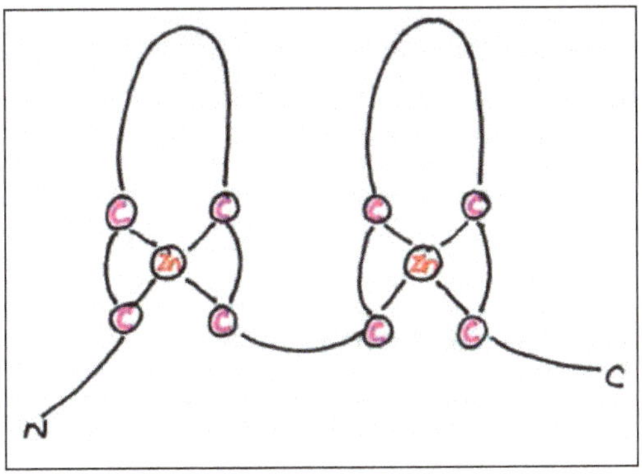

C_4 zinc finger motif.

Class III Zinc Finger Proteins (C_6)

Members of this class of zinc finger proteins possess the well-conserved motif Cys-X_2 –Cys-X_6 – Cys-$_{X5-12}$ –Cys-X_2 –Cys -X_{6-8} -Cys. The DNA binding domain has six cysteine residues bound to two zinc atoms. These proteins are also called as zinc clusters. They have three functional domains as shown in Figure below.

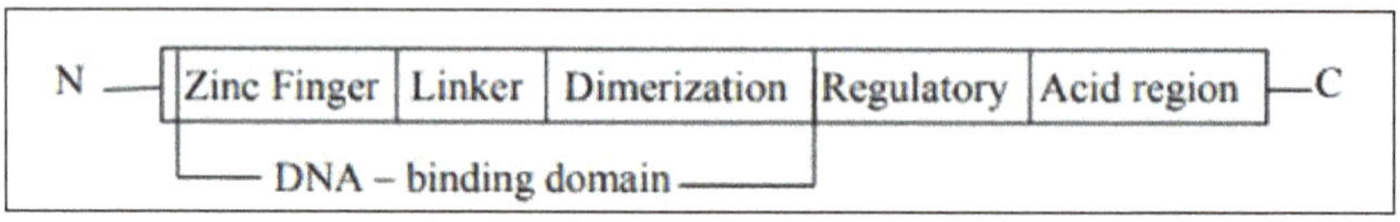

Functional domains of zinc cluster proteins.

The first and best studied zinc cluster protein is GAL4. It is a transcriptional activator of genes involved in the catabolism of galactose. The yeast gene regulatory protein GAL4 has the dinuclear $Zn_2(Cys)_6$ cluster. Its DNA-binding domain contains a conserved sequence of 6 cysteines (Cys- X_2-Cys-X_6-Cys-X_6-Cys-X_2-Cys-X_6-Cys). The two zinc coordinate to the six cysteines in a binuclear zinc thiolate cluster. The interatomic zinc-zinc distance is 3.5Å. Cys-11 and Cys-28 are the bridging ligands.

Zinc cluster proteins are also found in a variety of other fungal organisms such as Aspergillus nidulans, Kluyveromyces lactis, Schizosaccharomyces pomb and Candida albicans. The motif has not been observed in either prokaryotes or in higher eukaryotes.

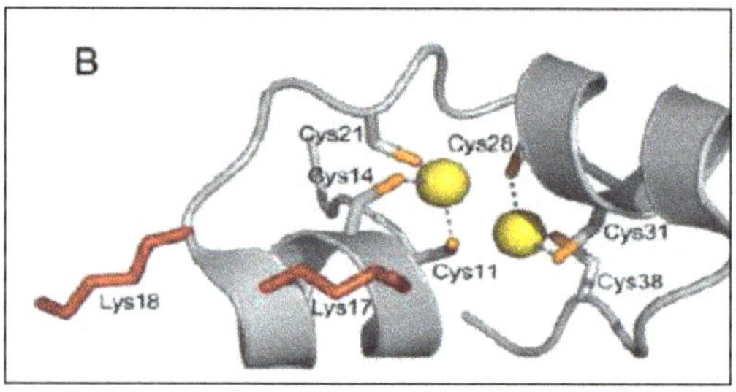

The Zn(II)2Cys_6 zinc cluster motif redraw.

More recently, a more systematic method has been used to classify zinc finger proteins. This method classifies zinc finger proteins into eight "fold groups" and the classification is based on the main chain conformation and secondary structure around the zinc-binding site. Each fold group constitutes proteins that share common structural features and are frequently functionally related. The different fold groups are: C_2H_2-like finger, Gag knuckle, treble clef finger, zinc ribbon, Zn_2/Cys_6-like finger, TAZ2 domain-like, short zincbinding loops, metallothioneins. The most common "fold groups" of zinc fingers are the Cys_2His_2-like (the "classic zinc finger"), treble clef, and zinc ribbon.

Biological Role of Zinc Finger Proteins

Zinc fingers are found in fungi, plants and animals. The typical number of finger repeats and the length of the linker between neighboring fingers varies greatly. Zinc finger proteins have diverse roles in many fundamental cellular processes such as transcription, translation, DNA replication and repair. Thus they participate in many biological functions such as metabolism, signaling, cell proliferation and apoptosis. They function as interaction modules that are involved in binding to DNA, RNA, proteins and small molecules like oxidizing agents, reducing agents, metals and therapeutic drugs.

In zinc finger proteins the role of zinc is purely structural. The main interactions between the protein and DNA are formed by the side chains of residues from a α-helix which generally bind to DNA at the major groove. Figure below shows a classic zinc finger (PDB entry 1znf) with 25 amino acids. Except the two cysteines, two histidines and the two hydrophobic amino acids phenylalanine and leucine the other amino acids face outwards.

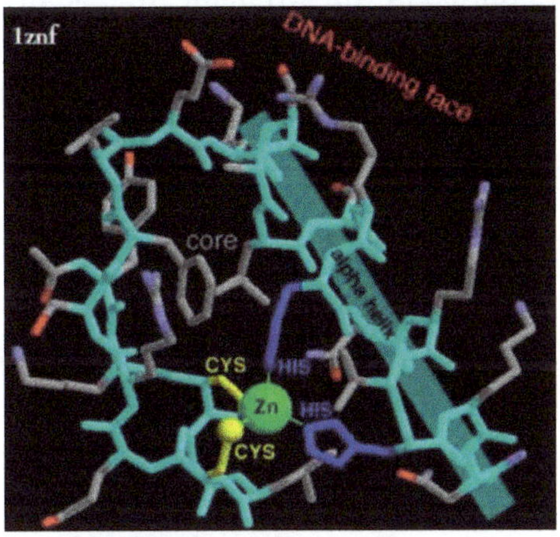

Classic zinc finger (PDB entry 1znf).

The top portion of Figure 7 shows the abovementioned classic zinc finger bound to DNA with three such fingers and the lower portion, the protein without the attached DNA.

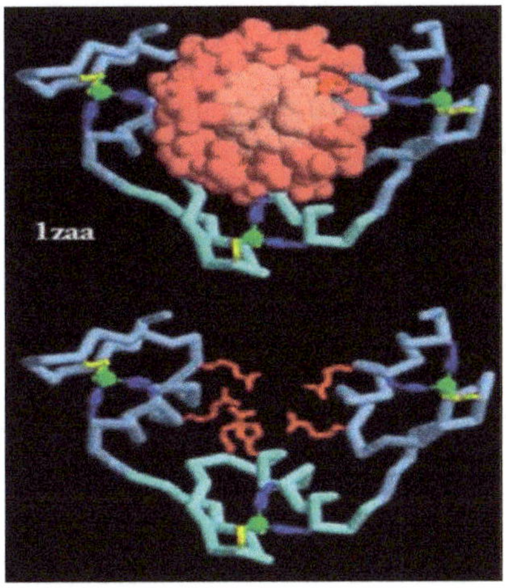

A zinc finger protein bound to DNA.

The interactions of zinc finger proteins with DNA can be understood using the C_2H_2

zinc fingers. C_2H_2 fingers bind DNA segments with a specific affinity because of the four amino acid residues present in the α-helix of the finger (in the N terminal part of the α-helix). These residues participate in DNA recognition by interacting with hydrogen donors and acceptors exposed in the major groove. The mode of binding of C_2H_2 finger to DNA was first determined by studying the protein Zif268.

This protein contains three zinc fingers, the α-helical portion of each finger fits in the major groove of DNA. The binding of successive fingers causes the protein to wrap around the DNA. Crystallographic analysis of zinc finger Zif268-DNA complex formed by interaction of three zinc fingers with DNA revealed that the three zinc fingers bind in an equivalent manner to a segment DNA nine base pairs long in the major groove so that each zinc finger contacts three base pairs with no gaps between the three sites. The linkers between the adjacent fingers play little or no role in DNA recognition.

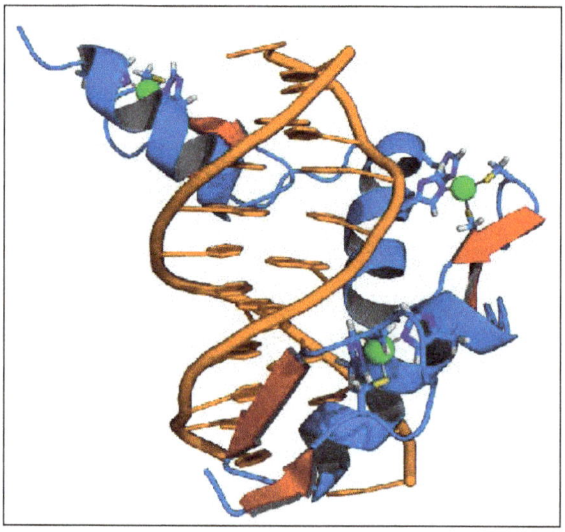

Structure of three fingers of zinc finger protein Zif268 bound to DNA (The protein shown in blue contains three zinc fingers in complex with DNA shown in orange. The zinc ion is represented as a green sphere coordinated.

Zinc Finger Engineering

In recent years considerable effort has been done in the field of designer zinc fingers. Designer zinc fingers are artificial zinc fingers (usually C_2H_2 zinc fingers) with different specificities. They can specifically target unique binding sites within the human genome and thus have important applications as probes and in human gene therapy. Efforts to increase the zinc finger DNA binding affinity can be done in two ways. One is to increase the number of zinc fingers. The second way is to change the linkers for connecting three finger modules. However, there is not much improvement in affinity when more than three fingers are connected by TGEKP linkers. So, more focus is on the latter approach.

References

- Structure-Function-Relationship-of-Metalloproteins, publication: ims.ac.jp, Retrieved 28, August 2020
- Iron-Sulfur-Proteins-and-their-Synthetic-Analogues-Structure-Reactivity-and-Redox-Properties: ias.ac.in, Retrieved 20, May 2020
- Classification-of-mononuclear-zinc-metal-sites-in-protein-structures-14231: pnas.org, Retrieved 03, March 2020
- Zinc-finger-proteins, chromatin-modifying-and-dna-binding-proteins, key-epigenetic-players: epigenie.com, Retrieved 22, July 2020

4

Biomineralization

The process through which mineral crystals are produced by living organisms is known as biomineralization. It involves nucleation and crystal growth. Some of the minerals which are produced through this process are calcium phosphate and calcium carbonate. This chapter discusses in detail the process of biomineralization and these minerals.

Biomineralization is the process by which living organisms orchestrate the synthesis and organization of minerals (biominerals), and it may be viewed as an ancient process for accumulation of metal ions in living systems. The structure and properties of biominerals have yet to be rivaled by any synthetic effort by scientists to date. Therefore, deciphering the assembly algorithms and the components that initiate and promote hierarchical deposition of cations has significant implications for the development of Nano composites and nanotechnology as a whole.

Many of us probably recollect walking along a sandy shore and marveling at the beautiful, ornate shells that are revealed and left behind by a retreating tide. The intricacy of the patterns on these shells are eye-catching, their texture and robustness a product of several million years of evolutionary engineering. They represent some of the marvels of materials conjured up by Mother Nature.

The primary constituent of the protective armor (mantle) that encases the soft body of mollusks and some bivalves is calcium carbonate. The mantle of a mollusk differs from that secreted by bivalves in that it is additionally composed of chitin, a biopolymer commonly associated with the exoskeleton of shrimps and lobsters. Biogenic calcite along with amorphous silica represents the two most common biominerals on the planet. Biomineralization is the assembly of these highly organized mineral phases by and in living systems and may be viewed as an accumulation of metal by an organism through a carefully orchestrated biological process, which, in some instances, has remained unchanged over millennia. Interestingly, biomineralization also provides a record of the environmental history of an organism over its lifetime.

The motivation to understand and shed light on biomineralization processes is multifaceted. The assembly of mineral phases achieved by living organisms is yet to be rivaled by any synthetic effort by scientists to date. Biominerals, due to their unique hierarchical organization, exhibit very interesting properties that have potential applications in all walks of life. An interesting aspect of biomineralization is that the formation of the mineral phase occurs from conditions that are not thermodynamically favorable for the mineral phase's nucleation and growth. Therefore, deciphering the assembly algorithm and components that initiate and promote hierarchical deposition of cations has significant implications for the development of nanocomposites and nanotechnology as a whole.

There are numerous examples of biomineralization and classes of biominerals in nature. This diversity provides a unique opportunity for creative and defining collaboration between materials scientists, chemists, biophysicists, and biologists for harnessing the structure–property relationships in these naturally occurring minerals. One can envisage applications for biominerals in several arenas ranging from military hardware, nanocontainers and nanosystems for medical and diagnostic applications, novel materials for photonics with hierarchical structures, and tunable photonic properties (bioinspired optics).

Role of Macromolecules in Biomineralization

Molecular systems recognized over the years as being important for biomineralization are well-defined highly charged proteins (biomineralization promoter proteins [BPPs]) and polysaccharides. Knowing the structure and chemical makeup of these macromolecules and the mechanism by which these molecules drive biomineralization can aid in the design of synthetic mimics.

Two common aspects have emerged with regard to macromolecules in biomineralization: (1) These macromolecules have a high degree of negative charge along their backbone, and (2) in the case of proteins, the presence of a β-sheet motif appears to be important. Both features are thought to aid in the initiation of crystal growth and possibly function as a molecular template for the organization of the mineral phase. In view of the compelling evidence for the part of proteins and polysaccharides in biomineralization, proteomics and glycomics can be expected to play an important role in furthering our knowledge base in the coming years.

Biogenic Calcite

Biogenic calcium carbonate occurs in two polymorphs: aragonite and biogenic calcite. Aragonite, which is a thermodynamically unstable form, will transform to calcite at elevated temperatures, although this transformation in nature is on the timescale of a few million years. Biogenic calcite is the primary constituent material of the mantles of mollusks and bivalves.

The formation of calcite occurs from a poorly organized precursor called amorphous calcium carbonate (ACC). Calcareous sponges and ascidians, which are invertebrate marine filter feeders and commonly referred to as sea squirts, synthesize skeletal elements composed of both ACC and calcite, with calcite covered in an ACC layer followed by a thinner calcite in the case of the former, and an ACC core enveloped in calcite in the case of the latter. This is interesting because calcareous sponges and ascidians are wide apart in the phylogenetic tree, in that calcareous sponges belong to the phylum Porifera, and, ascidians belong to the phylum Cordata. In both instances, the growth of these discrete phases was found be under the control of macromolecules. In the biomineralization of abalone shells, cooperative interactions between polyanionic proteins by themselves have been shown to promote the abrupt transition between calcite and c-axis, oriented aragonite.

Biominerals also play an important role in vertebrates. In avians, the eggshell is composed of calcium carbonate; two families of proteins, ovocleidins and ovocalyxins, are responsible for shell's biomineralization. The robust mechanical properties of the eggshell are due to the organized deposition of calcite spherulites on an organic framework. The eggshell is composed of six layers, with the calcified zone sandwiched between organic layers and organized in irregular cones.

Hydroxyapatite

Hydroxyapatite (HAP), a naturally occurring form of calcium phosphate (CP), is the mineral phase in bones that is responsible for conferring strength to the bony skeletal framework in vertebrates. Very often the ions within the HAP in bone are replaced by other ions, with the substitution of phosphate with carbonate being the most prevalent, resulting in the formation of carbonated HAP. A calcium deficient form of HAP is also found in tooth dentin and enamel. In bone, HAP is assembled within a fibrillar collagen network, with the triple helical collagen bundles (tropocollagen) forming a template of sorts for the nucleation and growth of the platelet-like HAP phase. While the origins of the platelet-like HAP phase in bone remain a work in progress, it is thought to occur through an octacalcium phosphate intermediate phase. Interestingly, the staggered arrangement (assembly) of adjacent and adjoining collagen bundles results in a 67 nm offset (or gap) overall. This gap is very critical for HAP formation and provides the vacancy in the organic framework for the interpenetrating deposition of collagen by the osteoblasts in bone and is also thought to be the nucleating site for HAP formation.

In addition to the collagen network and BPPs, the formation of HAP precursors can be influenced by other local variables. In a recent study, Researchers characterized the structure and organization within crystalline and biomimetic HAP using various characterization techniques, including cryo-transmission electron microscopy (TEM), and made a surprising find: water molecules can orient the formation of the amorphous apatite phase (similar to ACC), which covers the crystalline core in bone. This example not only emphasizes the need for robust characterization tools, but also highlights how much more still needs to be understood about biomineralization processes.

Bone is a spectacular example of an inorganic/organic nanocomposite; however, such nanocomposites are not a monopoly of vertebrates and inorganic–organic composites involving HAP are also common in invertebrates. The exceptional toughness of the dactyl clubbing appendage of the peacock mantis shrimp (Odontodactylus scyllarus) can be attributed to the presence of highly organized crystalline HAP, which in this case is present as a layered composite with chitin. Therefore, understanding the role of protein mediators in biogenic HAP formation within the bone environment has big implications for not only understanding pathologies that involve bone loss, such as osteoporosis, but also to enable the development of new coatings for improving bone conduction and induction and open up new treatment options for improving bone mineral density. Toward this larger objective, synthetic systems that can replicate the bone mineralization microenvironment could be important.

Magnetosomes

In addition to invertebrates and mammals, bacteria are also capable of processing minerals into highly ordered structures. Certain strains of marine bacteria, called magnetotactic bacteria, are capable of processing iron salts into magnetite. Researchers studied the size distribution and twinning of magnetite crystals from various strains of helical and rod-shaped marine magnetotactic bacteria (Magnetospirillum magnetotacticum, MC-1, MC-2, etc.) and observed that not only were the magnetite crystals produced by these magnetotactic bacteria of a much narrower size distribution than their abiotic cousins, they also exhibited an asymmetry in size distribution that is not observed in synthetic samples. TEM analysis of magnetosomes synthesized by magnetotactic bacterium MC-1 has revealed that control over crystal structure occurs at a very early stage of crystal growth, and the medium of the culture impacts both the morphology and number of magnetite crystals, with an acetate-containing medium yielding more crystals than a sulfide-containing medium. Such exquisite control over magnetite structure is of extreme relevance in many applications, including the development of nanotechnology-based therapeutics.

Biosilica

Biominerals composed of silica constitute the second most abundant mineral composition after biogenic calcium carbonate in nature. It is now well known that many plant species, such as rice, accumulate silica from the soil, and this accumulation is thought to have a beneficial effect for the survival of the plant. Diatoms constitute the most prevalent structure composed of amorphous silica found in nature. Diatoms are found in every body of water on the planet; each body of water has a unique population of diatoms, and this is frequently used in forensics to pinpoint the geographical location of the body of water. Diatoms are actually living systems: unicellular microalgae with well-defined exoskeletons of silica. The process of deposition of silica (i.e., silicification) occurs via silica deposition vesicles that are secreted by the microalgae. The factors and biological molecules involved in the nanofabrication and patterning of silica into intricate structures are just beginning to be understood.

The role of organic molecules in the biomineralization processes is now a well-understood phenomenon. Researchers identified and isolated a family of cationic polypeptides called silaffins from the diatom species Cylindrotheca fusiformis. Silaffins are tightly associated with the biosilica synthesized by the diatom, and it has been shown that silaffins and other polyamines play a direct role in biosynthesis of silica. The deposition of the silica skeleton is believed to occur through precipitation within the water phase, and the repeated phase separation that ensues to the sol-gel transition of the growing silica phase is thought to be responsible for the observed hierarchy in the diatom silica exoskeleton.

Diatoms offer an interesting platform to make highly functionalized silica structures. Reseachers have demonstrated that chemical modification of the diatom can be carried out without compromising its structure, via the incorporation of thiol moieties into the diatom during frustule synthesis. One can imagine such modified diatoms being used in drug delivery and catalysis.

Synthetic Systems

In spite of these recent advances in understanding the biomineralization process in bacteria, bivalves, and vertebrates, there are very few examples of synthetic systems that can replicate biomineralization processes. In fact, most efforts to date rely on protein extracts from the biomineralization system of interest to replicate the phase and structure of the biominerals. A notable example of a novel synthetic system that offers templated biomineralization is self-assembled peptide ampiphiles containing phosphorylated-serine residues capable of directing the mineralization of HAP such that its c-axis is aligned with the long axis of the self-assembled peptide fibers polyanionic proteins are known to mediate the formation of calcite and aragonite in sponges and abalone shells. However, in a recent study, researchers showed that control over the calcium carbonate phase can also be achieved using synthetic polymers. They studied the effect of poly(styrene sulfonate) (PSS) as a polyanionic mimic of biomineralization promoting proteins, in the formation of ACC. They found that the binding of calcium to PSS is a key step in the nucleation and growth of metastable ACC, thus providing more evidence that one of the primary roles of proteins and macromolecules in biomineralization might be in the nucleation and growth of metastable ACC, which is the precursor to calcite.

One direction of research could be the development of nanocomposites via direct biomineralization of nanoscale synthetic templates. In this context, researchers studied the formation of CP on silica nanoparticles (SNPs) bearing positive charges, collagen, and phosvitin. They found that generally while the formation of CP on SNPs was heterogeneous, negatively charged surfaces promoted a more uniform deposition of CP nodules, even in comparison to collagen-coated SNPs.

Together, these findings provide the foundation for developing synthetic systems wherein biomineralization can be realized in a controlled three-dimensional metal or polymer framework. In this context, synthetic biology offers tools and concepts to manipulate living systems at the genetic and protein expression level. Combined with optogenetics, this can provide a unique set of tools to controllably perturb the biomineralization environment within living systems and assess its impact on the structure of the mineral phase. Gaining control over the biomineralization process can foster the development of new materials for photonics and regenerative medicine, where hierarchical structures are critical for function.

Nucleation and Crystal Growth

Nucleation and crystal growth are processes that occur in supersaturated media and that have to be carefully controlled in a directed mineralization process. These requirements can be met in an organism through highly regulated active transport mechanisms and through specific modulation of the surface reactivity. The transport mechanisms may include transmembrane ion flux, ion (de)complexation, enzymatically catalyzed gas exchange (CO_2, O_2 and H_2S), local changes in redox potential ($Fe^{II}/^{III}$) or pH, and variations in the ionic strength of the medium. All these factors can create and sustain a supersaturated solution in a biological compartment. Nucleation, on the other hand, is linked to the kinetics of (heterogenous) surface reactions; processes such as cluster formation, anisotropic crystal growth and phase transformartions are also determined by the properties of the surface. In the biological realm there are also a number of surface structures that specifically prevent undesired nucleation; for instance, fishes in polar waters can thus protect themselves from ice formation in their bodily fluids at temperatures slightly below 0°C.

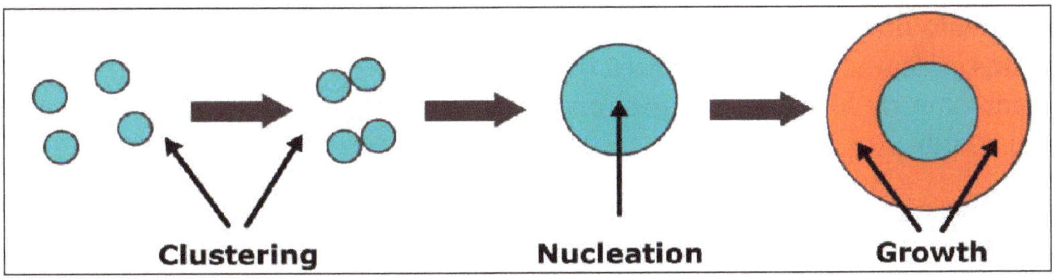

Nucleation and crystal growth.

The growth of a crystal or of an amorphous solid from the nucleus which is formed in an activation step (crystal nucleation energy) can proceed directly from the surrounding solution or by the continuous supplementation of the necessary ions or molecules. On the other hand, the diffusion of particles can be drastically altered through significant changes in the viscosity of the medium. e.g. through gel formation of a biological matrix – a mechanism which is presumably responsible for the deposition of amorphous silica in plants.

One of the most important and most fascinating aspects of biominerals is the fact that their shapes do not have to coincide at all with the regular crystallographic forms of the inorganic material. Far more significant for the final morphology are the spatial limitations through biopolymers, membranes or vesicles during the formation of biominerals, even if the completed structures are finally assembled outside such confinements. Of course, modifications of crystal growth may be induced nonbiologically; in fact, rather simple chemicals in the growth medium can influence the shape of the mineral formed. For example, the spindle shape of calcite crystals in gravity sensors can be reproduced by crystallization in the presence of 5 mM malonic acid, and monomolecular layers of stearic acid induce the crystallization of $CaCO_3$ as disc-shaped vaterite crystals rather than as rhombic calcite.

Calcium Phosphate

Biominerals are organic-inorganic composites, which fulfill various biological functions. In vertebrates, hard tissues provide body support, take part in tearing food, protect organs, and are reservoirs of calcium and phosphate. Understanding of molecular basis of biomineralization is essential to obtain new biomaterials. Biominerals are organic-inorganic composites, which fulfill various biological functions. In vertebrates, hard tissues provide body support, take part in tearing food, protect organs, and are reservoirs of calcium and phosphate. Understanding of molecular basis of biomineralization is essential to obtain new biomaterials.

Human hard tissues are formed of calcium orthophosphates. Among them, most important are amorphous calcium phosphate, octacalcium phosphate, calcium hydrogenphosphate dihydrate, and calcium-deficient apatite and hydroxyapatite (HA). In organisms, calcium orthophosphates occur mainly in the form of poorly crystallized nonstoichiometric sodium-, magnesium-, and carbonate-containing HA—so-called biological apatite.

Biomineralization is a multistep process, which requires using the structures of extracellular matrix vesicles, numerous enzymes and glycoproteins. Due to strict control, biominerals differ from pure chemical minerals. In contrast to geological minerals, biominerals are composite materials comprised of both inorganic and bioorganic components. The mineral constituent gives tissues hardness and resistance to mechanical damage. Stiffness of the tissue depends on the amount of inorganic components and organic phase. About 70% of bone tissue is made up of mineral structure, while the rest is water and organic substances. Moreover, having formed in vivo under well-controlled conditions, biomineral phases often have properties, such as shape, size, crystallinity, isotopic, and trace element compositions, quite unlike their inorganically formed counterparts.

The key factors, which determine the size, shape, internal structure and properties of biominerals, are proteins, which control the nucleation and growth of the crystals. Biomineralization involves protein-protein interactions and interactions between proteins and inorganic fraction. Among them, two major groups can be distinguished. Scaffold for a growing mineral phase provides insoluble collagen. In human bones, collagen makes up to 20–30% wt. Nucleation and crystal growth regulation are controlled by soluble, noncollagenous proteins. They are usually highly acidic, undergo extensive posttranslational modifications, and frequently belong to the group of intrinsically disordered proteins (IDPs). IDPs are dynamic, flexible, heterogeneous populations of molecules without a well-defined folded structure. The highly charged character, along with the low content of hydrophobic amino acid residues, results in strong electrostatic repulsion and the lack of a well-packed hydrophobic core. High content of IDPs' carboxyl and phosphate acidic groups involved in biomineralization results in high calcium binding capabilities. Examples of proteins engaged in bones and teeth biomineralization are shown in Table.

Table: Proteins involved in bone and teeth biomineralization.

Protein name	Accession number	pI	Mass (kDa)	Overall percent disordered—PONDR
Osteopontin	AAA59974.1	4.4	~34	70%
Bone sialoprotein	AAA60549.1	4.1	~35	59%
Matrix extracellular phosphoglycoprotein	AAK70343.1	8.6	~58	64%
Dentin matrix protein-1	AAC51332.1	4.0	~56	90%
Dentin sialophosphoprotein (preproprotein)	NP_055023.2	3.6	~131	88%
Ameloblastin	AAG35772.1	4.9	~48	58%
Amelogenin	AAC21581.1	6.5	~22	65%

The Basic Division of Proteins Involved in the Formation of Calcium Phosphate Biominerals

Bone and teeth biomineralization takes place within the extracellular matrix (ECM), outside the cells. ECM is a highly organized and complex structure, unique to the specific organ type. It is an organic, noncellular fraction of mineralized tissues, composed of structural and functional proteins. The content of the organic fraction varies depending on the type of tissue and constitutes 30% of bone, 20–25% of dentin, and only 0.5% of mature enamel. The basic division of proteins involved in the formation of calcium phosphate biominerals includes insoluble matrix like collagen and soluble noncollagenous proteins (NCPs).

Collagenous Proteins of ECM

Among organic phase components, there are insoluble substances, which include structural macromolecules, creating a scaffold for the growing mineral. They give shape to the resulting crystal and can act as initiators for the creation of crystallization roots. An example of such substance is collagen. Animal tissues primarily contain collagen type I or collagen type II. Collagen is the major ECM molecule that self-assembles into cross-striated fibrils, provides support for cell growth, and is responsible for the mechanical resilience of connective tissues. The term “collagen” defines a whole family of glycoproteins. The most common motifs in the amino acid sequence of collagen are repeating sequence [Gly-X-Y]n, both with and without interruptions. The X and Y positions are occupied by proline and its hydroxylated form, hydroxyproline, respectively. The right-handed triple helix is formed from three left-handed polyproline α-chains of identical length, which gives collagen a unique quaternary structure. Hydroxylation of the rest of the proline in collagen is necessary for the stabilization of the triple helix. Collagen chains create interactions leading to the organization of chains in a four-stage array. Collagen fibers are also cross-linked via lysyl residues. Electron microscopy techniques allow to observe successive regions of overlap and gaps, resulting from the mutual arrangement of collagen fibers.

Noncollagenous Proteins (NCPs) of Extracellular Matrix (ECM)

While structural collagen is the main protein of ECM, NCPs are important for the regulation of biomineralization. NCPs are usually acidic, capable of binding large amounts of calcium ions. Proteins, which directly influence biomineralization, act as nucleators and regulators of crystal growth and orientation. NCPs are frequently posttranslationally modified, e.g., phosphorylated, glycosylated, or proteolytically processed. A large number of noncollagenous proteins have disordered secondary and tertiary structures and belong to a group of intrinsically disordered proteins.

Intrinsically Disordered Proteins in Calcium Phosphate Biomineralization

IDPs are a group of proteins that have gained a special interest of researchers for over 20 years. Their discovery was especially challenging for traditional protein structure paradigm, stating that protein function depends on a fixed tertiary structure. The IDP group includes proteins with altered secondary and tertiary structures from total random coil structure to some intrinsically disordered regions. They are characterized by considerable plasticity, flexibility, and high conformational dynamics. IDP features are high net charge and low hydrophobicity, which is not conducive to the formation of the hydrophobic core. They often bind low molecular ligands and macromolecules such as ions and proteins. By interacting with ligands, IDPs can undergo local or global structuring. They can also fulfill their functions while remaining completely disordered. Other characteristics of this group are multiple amino acid repeats and susceptibility to posttranslational modifications. An open structure and lack of a packed core increase the availability of potential phosphorylation, glycosylation, and proteolytic cleavage sites.

Extended, flexible structure, highly acidic nature, susceptibility to modifications and especially the ability to interact with many different partners make these proteins particularly adapted to the biomineralization control. Many NCPs belong to IDP group. Family of small integrin-binding ligand N-linked glycoprotein (SIBLING) is the example of IDPs engaged in HA formation and is presented below. In addition to SIBLINGs, enamel matrix proteins such as ameloblastin, amelotin, and enamelin also have disordered structures. More information about the proteins involved in HA biomineralization belonging to IDPs can be found in reviews.

Siblings

An example of NCPs is the family of small integrin-binding ligand N-linked glycoprotein (SIBLING). It is a group of proteins identified in bone, dentine, and cementum, which includes osteopontin (OPN), bone sialoprotein (BSP), dentin matrix protein-1 (DMP1), dentin sialophosphoprotein (DSPP), and matrix extracellular phosphoglycoprotein (MEPE). It is believed that they all evolved as a result of gene duplication from extracellular calcium-binding phosphoprotein family. Although proteins differ in composition of amino-acid sequence, SIBLINGs share common features. Their genes are located in a 375 kb region on the 4q21 human and 5q mouse chromosomes. They display a similar exon structure. Their sequence contains an Arg-Gly-Asp (RGD) motif, which mediates cell attachment/signaling by binding to cell-surface integrins. SIBLINGs belong to IDP group and are highly acidic. They share numerous sequence repeats, which are often observed in the case of IDPs. Particularly important for their biomineralization activity may be acidic serine- and aspartate-rich motifs (ASARM), which are involved in the phosphate administration [29, 30]. SILBINGs are frequently posttranslationally medicated. Some of their biological functions depend on phosphorylation, glycosylation, proteolytic processing, sulfonation, or transglutaminase cross linking.

Osteopontin

Osteopontin (OPN), also known as secreted phosphoprotein 1 (SSP_1), is a highly phosphorylated and glycosylated sialoprotein. It is a multifunctional protein expressed by several cell types including osteoblasts, osteocytes, as well as osteoclasts or odontoblasts. OPN undergoes many posttranslational modifications such as phosphorylation, glycosylation, sulfonation, or proteolytic processing, and modifications vary depending on a protein role and localization. In bones and dentin, osteopontin is located at the site of biomineral formation. Synthesis of OPN is stimulated by calcitriol, physiologically active form of vitamin D, which is known as a trigger for bone destruction and remodeling.

Osteopontin was originally identified as a bridge between the cell and HA in ECM of bone. It is known that OPN increases the adhesion of bone cells by concentrating in mineralized collagen matrix during bone formation and remodeling. The protein is highly produced by developing osteoblasts and osteoclasts, and it has been shown to regulate both cell type adhesions.

Another role of ONP during biomineralization is modulation of osteoclastic function. OPN binding to integrin $\alpha v\beta_3$ is essential for regulation of osteoclastic activity and is necessary in the formation of a sealing zone. Through CD44-associated cell signaling, OPN stimulates osteoclast migration.

As a highly acidic phosphoprotein, OPN binds to the surface of hydroxyapatite crystals through the electrostatic interactions between crystals and carboxyl and phosphate groups. Furthermore, disordered structure of OPN can promote the binding of calcium ions. The results of in vitro and in vivo studies suggest that osteopontin has an inhibitory effect on HA formation and growth. OPN knockout mice display increased mineral content and size. Therefore, it is suggested that OPN is one of the proteins whose role is to prevent crystal formation in soft tissues. Peptides of OPN obtained after proteolytic processing differ in a biomineralization effect. ASARM peptide of OPN binds to HA crystals and consequently inhibits ECM matrix mineralization. Its inhibition activity is dependent on serine residues' phosphorylation. N- and C-terminal fragments of OPN from milk promote de novo HA formation, but at the same time, the central fragment showed an inhibitory effect.

Bone Sialoprotein

Bone sialoprotein (BSP) is one of the most abundant NCPs of bone. In contrast to OPN, BSP is only localized in tissues that undergo mineralization: bone, dentin and mineralizing cartilage, and cementum. The protein is produced by osteoblasts, osteocytes, osteoclasts, as well as by chondrocytes. Disordered structure of BSP was shown by NMR, CD, and SAXS studies. BSP is highly glycosylated especially at C-terminal fragment, and to a lesser extent also phosphorylated and sulfonated. N-terminal region of the protein contains collagen-binding motif. In vivo and in vitro studies indicated that BSP can be involved in initial stages of hydroxyapatite biomineralization. BSP is intensely expressed by osteoblastic cells in sites of primary bone formation. The protein is present at mineralizing boundaries of bone, dentin, and calcifying cartilage tissues. On the other hand, BSP increases osteoclastogenesis, and in that way, it can initiate bone remodeling. BSP knockout mice are characterized by short, hypomineralized bones with high trabecular bone mass. Overexpression of BSP leads to dwarfism, decreased bone mineral density, and decreased trabecular bone volume.

It was shown that BSP has high affinity to hydroxyapatite and acts as a de novo nucleator of HA crystals in vitro. The protein also binds to type I collagen, and by interaction with collagen, BSP could regulate HA nucleation. Moreover, control of mineral formation is highly associated with the state of phosphorylation, sulfonation, and glycosylation of BSP.

When creating skull bones, including the mandible, alveoli, and skull vault, the osteoprogenitor cells secrete a collagen extracellular matrix—osteoid. Skull bones are formed by intracerebral ossification. BSP affects biomineralization of the matrix in the ossification process, hence, the formation, shaping, and growth of hydroxyapatite crystals. In the absence of BSP, bone formation is delayed and osteoblast activity is impaired.

Matrix Extracellular Phosphoglycoprotein

Matrix extracellular phosphoglycoprotein (MEPE) is located in mineralized tissues as bone and dentin, but was found also in nonmineralized organs. The protein is primarily expressed by osteoblasts and osteocytes that are embedded within the mineralized matrix in bone and by odontoblasts during odontogenesis. Similar to osteopontin, MEPE seems to be a multifunctional ECM protein. Results of in vivo and in vitro studies suggest that MEPE is important for bone mineralization, Pi homeostasis and cell attachment. The protein was originally identified as interacting with PHEX. PHEX by binding to MEPE protects the protein from proteolytic cleavage by cathepsin B. Cleavage by cathepsin B leads to the release of the ASRAM peptide. The ASRAM motif is a functional domain of MEPE responsible for its inhibitory activity. The peptide may be responsible for phosphate and calcium regulation during the mineralization process. ASARM peptide inhibits hydroxyapatite mineralization by binding free calcium with high avidity, and their inhibitory effect is highly dependent on serine residues phosphorylation. AC100, another MEPE fragment, containing the integrin binding (RGD) and glycosaminoglycan-attachment (SGDG) motifs stimulates new bone formation in vitro and in vivo.

Dentin Matrix Protein-1

Dentin matrix protein-1 (DMP1) was the first molecule identified by cloning from the dentin matrix. Besides dentin, the protein is located also in bone and cementum as well as in other nonmineralized tissues. It is remarkably acidic and is a hydrophilic protein with many serine, aspartate, and glutamate residues. The protein is characterized by disordered structure that can aggregate in the presence of calcium ions.

It was shown that DMP1 is essential for both odontoblasts and osteoblasts maturation. The protein controls odontogenesis, osteogenesis, and Pi homeostasis. The effect of DMP1 on HA biomineralization depends on posttranslational modifications. It was shown that phosphorylated DMP1 inhibits HA formation and growth, while the dephosphorylated form acts as a HA nucleator.

It was demonstrated that DMP1 is proteolytically processed into N-terminal 37-K fragment and 57-K fragment from the C-terminal region, and it is likely that full-length DMP1 is a precursor to these functional fragments. It seems that the 57-K fragment plays a key role in the biomineralization process. The 57-K fragment contains 41 phosphate groups, while the 37-K fragment has only 12 phosphate groups. In addition, the 57-K fragment contains many functional sequences and domains such as the RGD motif, the ASARM peptide, and the peptide functioning as nucleator. In contrast to the full-length DMP1, the 57-K fragment promoted HA nucleation and growth. Furthermore, it was shown that DMP1 N-terminal fragment is able to stabilize the metastable amorphous calcium phosphate. Due to high aspartic acid residual content, the 37-K fragment can bind calcium ions very strongly favoring formation and stabilization of the amorphous nuclei. It was shown that the DMP1 57-K fragment also controls calcium carbonate mineralization in vitro.

Dentin Sialophosphoprotein

Dentin sialophosphoprotein (DSPP) was discovered by cDNA cloning using a mouse odontoblast cDNA library and was the first believed to be connected only with dentin. Subsequent studies have revealed that the protein is expressed also in bone, cementum, and some nonmineralized tissues. Analysis of DSPP knockout mice and mutation in the DSPP gene indicated a special role of protein in dentin mineralization. In humans, a mutation in the DSPP gene causes dentin hypomineralization and significant tooth decay, named dentinogenesis imperfecta. Additionally, studies of DSPP-null mice suggest that DSPP is crucial in the initial mineralization of bone as well as in the remodeling of the skeleton and therefore on bone turnover.

It is suggested that DSPP is a precursor protein activated after proteolytic processing. Cleavage of DSPP by zinc metalloproteinase bone morphogenetic protein-1 (BPM1) results in two protein fragments: dentin phosphoprotein (phosphophoryn) (DPP) and dentin sialoprotein (DSP). Subsequently, cleavage of C-terminal of DSP by matrix metalloproteinase 2 (MMP2) and MMP20 leads to the release of the third fragment called dentin glycoprotein (DGP), which has a strong affinity to hydroxyapatite.

Highly acidic DPP is the most abundant NCP of ECM in dentin. Amino acid sequence of DPP is composed mainly of aspartic acid and serine residues, whereas ~80% of serine residues are phosphorylated and phosphorylation is crucial to its function. The protein is also glycosylated and contains ASARM peptide. Due to its amino acid composition and phosphorylation, DPP binds large amounts of calcium ions with high affinity. The protein is involved in nucleation and control of the formation and growth of HA crystals during dentin mineralization [82, 83]. DPP binds to collagen fibrils and is present in front of dentin mineralization [84, 85]. There is a hypothesis that in front of mineralization, DPP is promoting the formation of initial apatite crystals. When predentin is converted to dentin, DPP with other proteins binds to the growing HA faces and inhibits or slows down crystal growth.

Tooth

Dentin, enamel, and cementum are calcified tissues and are major components of teeth. Dentin is usually covered by enamel on the crown and cementum on the root and surrounds the entire pulp. About 70% (wt) of dentin consists of the mineral phase, 20% (wt) is an organic material, and 10% (wt) is water. Enamel is the hardest substance in the human body and contains 96% (wt) of mineral phase. Enamel and dentin are created by two layers, namely odontoblasts and ameloblasts. Dentin is characterized by a high content of collagen I comprising ~90% of the dentin organic matrix.

Enamel

The organic part of enamel consists mainly of hydrophobic proteins, known as amelogenins and anionic proteins, which include ameloblastin, enamelin, and tuftelin. Enamel also contains sialophosphoproteins and enamel proteases such as matrix-20 metalloproteinase (MMP-20, matrix metalloproteinase-20, also known as enamelysin) and kallikrein 4 (KLK4). At the end of the crystallization, enamel loses almost entire organic matrix, which is degraded and replaced by hydroxyapatite crystals. Residues, small peptides, and amino acids account for only 1% of enamel and do not resemble the original matrix.

Amelogenins are a group of heterogeneous proteins present in odontoblasts and identified in enamel of maturing teeth. They are the main proteins of the organic enamel matrix, constituting over 90% of the pool of proteins secreted by ameloblasts into the intercellular matrix during the formation of enamel. These are hydrophobic proteins rich in proline, glutamine, leucine, and histidine. Amelogenins isolated from various organisms are characterized by high homology. Amelogenins belong to IDP groups; conformational analysis by CD and NMR spectroscopy showed that recombinant porcine amelogenin rP172 exists in an extended, unfolded state in the monomeric form. The extended, labile conformation of rP172 amelogenin is compatible with known functions of amelogenin in enamel biomineralization, i.e., self-assembly, associations with other enamel matrix proteins and with calcium phosphate biominerals, and interaction with cell receptors.

Amelogenin controls the organization and growth of enamel crystals, and its presence is critical for normal enamel formation. Defects in amelogenin sequence lead to defective enamel crystal formation and organization.

Ameloblastin, also known as amelin, is a tooth-specific glycoprotein. Unlike amelogenin, ameloblastin is located close to the cell surface. Ameloblastin accounts for about 5–10% of all proteins present in the enamel, and it is the second most abundant protein among the intercellular matrix of the enamel. Ameloblastin is also synthesized in dentin and cement, but its role in these tissues is not determined. Ameloblastin is a molecule of cell adhesion required to maintain a diversified state of ameloblasts. Bioinformatic analyzes and molecular modeling of the protein structure suggests that it may belong to IDPs, which was confirmed by the CD spectra of recombinant mouse protein.

The N-terminal fragment of ameloblastin comprising the self-assembly motif was shown to colocalize with amelogenin across the entire growing enamel, indicating the role of the two proteins in the organization of the linear growth of HA crystallites. Both proteins self-assemble into higher order structures from monomeric subunits, similar to type I collagen, the predominant matrix protein of bones, and dentin. Full-length amelogenin undergoes hierarchical stepwise assembly, first forming oligomers, which in turn assemble into higher order structures and stabilize mineral prenucleation clusters, and organize them into parallel arrays of linear chains, yielding the formation of crystallite bundles. There is a hypothesis that the higher order structures of the self-assembled ameloblastin (or most likely its N-terminal moiety) contribute to the oriented growth of the linear chains of amelogenin in the 3D space.

Dentin

Dentin forms a support for hard tissues of a tooth. It covers both crown and root structures and is responsible for its shape. The formation of dentin is closely related to bone. Both bone and dentin are composed of a collagenous matrix and a mineral phase with hydroxyapatite plate-like crystals. Mineralization of dentin and bone extracellular matrix is initiated with the aid of matrix vesicles and later involves secretion of families of a specialized matrix protein. In contrast to bones, dentin is not remodeled and does not participate in the regulation of calcium and phosphate metabolism. The tooth formation takes place throughout the lifetime of a tooth. Unlike enamel, dentin mineral content increases with age due to continuous mineral deposition either as physiological secondary dentin or as tertiary dentin following injury.

Dentin is composed of a collagen I, which accounts for about 92% of organic components. In addition to collagen, there is also a group of noncollagen proteins, which include dentin-specific phosphoryls (DPPs) and dentin sialoproteins (DSP), sialic acid, osteopontin, dentin matrix proteins (DMP1, DMP2, DMP3), BSP, acidic bone-75 glycoprotein, osteonectin, and proteins rich in γ-carboxyglutamic acid. Among organic components should also be mentioned proteoglycans, growth factors, phospholipids, and enzymes. Inorganic dentin components make up to about 70% of the tissue mass, making the dentin harder than the bone. The mass of a single hydroxyapatite crystal is about 10 times greater than in bone, but many times less than in enamel, and its size is about $35 \times 10 \times 100$ nm.

Cementum

Cementum protects the dental root dentin with a very thin layer. In many respects, it is very unique: it is not vascularized and is not innervated, it does not undergo constant remodeling like a bone, but it grows all the time. Cellular and acellular cement are distinguished based on the presence of cementocytes in its structure. The structure and composition of cement resembles bone tissue. Inorganic constituents make up to about 65% of tissue mass and consist mainly of hydroxyapatite crystals.

Cementum provides contact between roots of the teeth and periodontium ligaments. hrCEMP1 (recombinant human cementum protein 1) is an isolated form of cementum human β-sheet protein. It has been showed that hrCEMP1 forms clusters of 6.5 nm diameter. hrCEMP1 is an inducer of specific biomineralization proteins. It stimulates the differentiation of osteoblasts and cementoblasts. It is suggested that hrCEMP1 plays a significant role in octacalcium phosphate biomineralization and helps its binding to hydroxyapatite even without posttranslational modifications. X-ray diffraction measurements showed that hrCEMP1 is an inducer of polymorphic crystals.

Bone

The matrix of bone consists of type I collagen molecule self-assembled into a triple helix consisting of two α1 and one α2 chains (ca. 300 nm in length and 1.5 nm in diameter) and hydroxyapatite (Ca10(PO4)6(OH)2) nanocrystals (plate-shaped, 1.5–4 nm in thickness) deposited along collagen fibrils. The collagen fiber does not undergo spontaneous mineralization in the presence of phosphate and calcium ions, and the involvement of noncollagen proteins is necessary. The majority of these proteins belong to the SIBLING family described above.

Bones and dentin are characterized by a similar composition and the mechanism of their formation. Their organic matrix consists of type I collagen and the mineral matrix built of hydroxyapatite. Osteoblasts and odontoblasts, cells involved in osteogenesis and dentinogenesis, first secrete a nonmineralized matrix—a bone osteoid, and in case of dentin—predentin and then with fibers of collagen I form matrix for biomineralization. Calcium and phosphate ions are dislocated from the vascular network into the mineralization matrix. Osteoid and predentin contain numerous noncollagenous proteins called NCPs. On the basis of mutation experiments and suppression of NCP genes, it is suggested that they stimulate the nucleation and growth of hydroxyapatite crystals.

Bones and dentin differ significantly. Bone is a dynamic tissue because it is constantly remodeled, while dentin is a rather static tissue. Osteoblasts produce matrix components as a result of controlling growth factors to form a bone. Hydroxyapatite crystals grow above vesicles, thus creating mineralization centers. Bones have a lot of vessels, and they store calcium ions. The reason for bone resorption is the action of lysosomal enzymes produced by osteoclasts and lowering the pH in extracellular bone matrix.

Osteoclasts are responsible for the attachment, bone degradation, and subsequent tissue resorption. It was found that this is possible due to the strong polarization of osteoclasts. They have both frontal and posterior abdominal polarization. On the underside of the osteoclasts, there are structures responsible for the degradation of the mineral surface. The adhesion zone is located in the sealing zone—a ring made of actin and podosomes in which extracellular resorption pit is segregated and is used to retain heavy substances (minerals) formed during bone resorption near the bone surface. Podosomes are dynamic structures built of actin. They perform important functions, including adhesion, destruction of the bone and its matrix, and recognition of the appropriate medium. Podosomes contain a large number of V-ATPase molecules that pump protons and secrete proteases, which promote bone degradation.

Calcium Carbonate

Minerals known to be formed via biologically induced mineralization through passive surface-mediated mineralization include Fe, Mn, and other metal oxides, e.g., ferrihydrite ($5Fe_2O_3{\cdot}9H_2O$), hematite (α-Fe_2O_3), and goethite (α-FeOOH); metal sulfates, phosphates, and carbonates; phosphorite; Fe and Fe-Al silicates; and metal sulfides. Of all the minerals that have been associated with bio mineralization, carbonates are the most obvious. Microbially induced calcium carbonate precipitation (MICCP), most widely studied branch of biomineralization holds promise for variety of fields ranging from Biotechnology, Geotechnology, and Paleobiology to Civil Engineering. It has implications for:

- Atmospheric CO_2 fixation through carbonate sediment formation and lithification and dolomite precipitation.
- Solid-phase capture of inorganic contaminants.
- Pathological formation of mineral concretions, such as gallstones and kidney stones in humans.
- The possibility of understanding extraterrestrial biological processes such as Martian carbonate production by bacteria.

Microbially Induced $Caco_3$ Precipitation

Calcium carbonate precipitation is a rather straightforward chemical process governed mainly by four key factors:

- The calcium concentration.
- The concentration of dissolved inorganic carbon (DIC).
- The pH.
- The availability of nucleation sites. $CaCO_3$ precipitation requires sufficient calcium and carbonate ions so that the ion activity product (IAP) exceeds the solubility constant (K_{so}).

From the comparison of the IAP with the K_{so}, the saturation state (Ω) of the system can be defined: if $\Omega > 1$ the system is oversaturated and precipitation is likely as:

$$Ca^{2+} + CO^{\overline{2-}}{}^{3} \leftrightarrow CaCO_3$$

$$\Omega = a\left(C_a^{\;2+}\right)a\left(CO\frac{2-}{3}\right)/K_{so} \quad \text{With} \quad K_{socalcite,25^0}$$

$$= 4.8 \times 10^{-9}$$

The concentration of carbonate ions is related to the concentration of DIC and the pH of a given aquatic system. In addition, the concentration of DIC depends on several environmental parameters such as temperature and the partial pressure of carbon dioxide (for systems exposed to the atmosphere). The equilibrium reactions and constants governing the dissolution of CO2 in aqueous media (25°C and 1 atm) are given in Equation:

$$CO_{2(g)} \leftrightarrow CO_{2(aq.)} (pK_H = 1.468)$$

$$CO_{2(aq.)} + H_2O \leftrightarrow H_2CO_3 * (pK = 2.84)$$

$$H_2CO_{3*} \leftrightarrow H^+ + HCO_3^- (pK1 = 6.352)$$

$$HCO_3^- \leftrightarrow CO_3^{2-} + H^+ (pK2 = 10.329)$$

With $H_2CO_{3*} = CO_{2(aq.)} + H_2CO_3$

Under natural conditions, the precipitation of carbonates occurs very slowly over long geological times but in order to produce large amounts of carbonates shortly there is need to look for microbes with the ability to create conditions for precipitation of carbonates in shorter times. Different bacterial species precipitate carbonates in alkaline environments rich in Ca^{2+} ions and various mechanisms which could induce precipitation by bacteria in natural habitats have been proposed. Though, the precise role of bacteria and bacterial activities in carbonate crystallization remains unclear, they seem to fall in 3 categories:

- As per the first hypothesis, mineralization occurs as a byproduct of microbial metabolism.
- As per the second hypothesis, carbonate nucleation takes place on the cell wall.
- Lastly, the third hypothesis involves role of extracellular macromolecules.

Mainly four groups of microorganisms are seen to be involved in the process:

- Photosynthetic organisms—such as cyanobacteria and algae.
- Sulphates reducing bacteria—those are responsible for dissimilatory reduction of sulphates.
- Organisms utilizing organic acids.
- Organisms that are involved in the nitrogen cycle either ammonification of amino acids/ nitrate reduction/ hydrolysis of urea.

Microbially induced calcium carbonate precipitation (MICCP) via urea hydrolysis is the simplest and most widely used method for precipitation of carbonates for several technical applications.

MICCP via Urea Hydrolysis

The precipitation of carbonates via urea hydrolysis by ureolytic bacteria is the most straightforward and most easily controlled mechanism of MICCP with potential to produce high amounts of carbonates in short period of time. During microbial urease activity, 1 mol of urea is hydrolyzed intracellularly to 1 mol of ammonia and 1 mol of carbonate, which spontaneously hydrolyzes to form additional 1 mol of ammonia and carbonic acid as follows:

$$CO(NH_2)_2 + H_2O \xrightarrow{bacteria} NH_2COOH + NH_3$$

$$NH_2COOH + H_2O \rightarrow NH_3 + H_2CO_3$$

These products equilibrate in water to form bicarbonate, 1 mol of ammonium and hydroxide ions which give rise to pH increase.

$$H_2CO_3 \rightarrow 2H^+ + 2CO_3^{2-}$$

$$NH_3 + H_2O \rightarrow NH^{4+} + OH^-$$

$$Ca^{2+} + CO_3^{2-} \rightarrow CaCO_3 (K_{SP} = 3.8 \times 10^{-9})$$

KSP is the solubility product in $Ca^{2+} + CO_3^{2-} \rightarrow CaCO_3 (K_{SP} = 3.8 \times 10^{-9})$

The primary role of bacteria has been ascribed to their ability to create an alkaline environment through various physiological activities.. Bacterial surfaces also play an important role in calcium precipitation. Due to the presence of several negatively charged groups, at a neutral pH, positively charged metal ions could be bound on bacterial surfaces, favoring heterogeneous nucleation. Commonly, carbonate precipitates develop on the external surface of bacterial cells by successive stratification and bacteria can be embedded in growing carbonate crystals.

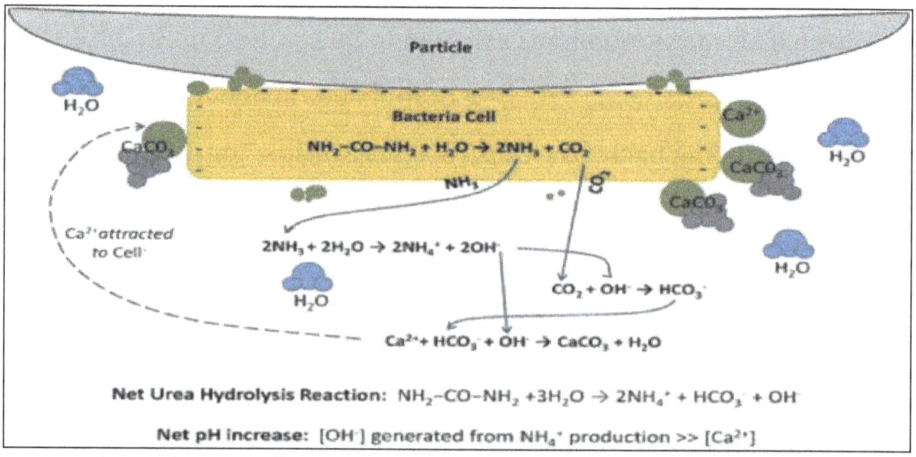

Bacteria serving as nucleation site for CaCO3 precipitation in the sand particles.

Possible biochemical reactions in urea-CaCl2 medium to precipitate CaCO3 at the cell surface can be summarized as follows:

$$Ca^{2+} + Cell \rightarrow Cell - Ca^{2+}$$

$$Cl^{-} + HCO^{3-} + NH_3 \rightarrow NH_4Cl + CO_3^{2-}$$

$$Cell - Ca^{2+} + CO_3^{2-} \rightarrow Cell - CaCO_3$$

Enzyme Urease

Several authors have investigated various conditions for calcium carbonate precipitation via urea hydrolysis. The actual role of the bacteria in the precipitation remains, however, a matter of debate. Some authors believe this precipitation to be an unwanted and accidental by-product of the metabolism while others think that it is a specific process with ecological benefits for the precipitating organisms. These bacteria include Bacillus pasteurii, Pseudomonas sp., Variovorax sp., Leuconostoc mesenteroides, Micrococcus sp., Bacillus subtilis, Deleya halophila, Halomonas eurihalina and Myxococcus xanthus.

Reaction Conditions Reported in the Literature for Production of $CaCO_3$ Via Urea Hydrolysis

Purpose	Urea (mM)	Ca^{2+} (mM)	Urease activity (mM/min)
Sr^{90} sequestration	333	25	0.045
Sr^{90} sequestration	330	0.025	0.042
Removal of Ca^{2+} from waste water	16	14	0.293
Removal of Ca^{2+} from waste water	8	15	0.032
Stone remediation	333	12–50	0.110
	333	340	0.02–0.12
Stone remediation	66	25	0.041
Portland cement remediation	333	50	n/s
Plugging of rock pores	333	0.25	n/s
Bio cementation	1500	1500	4–18
Sand column cementation	333	25	0.65
Soil cementation	1000	1000	n/s
Bio deposition	333	340	n/s
Carbonate precipitation	666	250	n/s

Polymorphism of Carbonate Crystals

Studies have reported that bacterial induced calcium carbonate precipitation results in the production of different phases of $CaCO_3$. Calcium carbonate forms three anhydrous polymorphs: calcite, aragonite and vaterite, two hydrated crystalline phases: monohydrocalcite ($CaCO_3 \cdot H_2O$) and ikaite ($CaCO_3 \cdot 6H_2O$), and various amorphous phases (ACC) with differences in short range order and degree of hydration. Although vaterite and calcite are the most common bacterial calcium carbonate polymorphs, mineralization of monohydrocalcite and aragonite) have also been reported. The evidence that bacterial mineralization of calcium carbonate involves the formation of ACC precursor phases is also growing. The species specific precipitation of carbonate bio minerals by various bacterial isolates has also been reported by many. But despite extensive studies on bacterial carbonatogenesis, little is known on what is the cause(s) of polymorph selection during bacterial calcium carbonate mineralization. Studies suggested that phase, amount and morphology of calcium carbonate minerals depend on supersaturating, temperature, pH and $[Ca^{2+}] / [CO^{-32}]$ ratio. The saturation index, SI = log Ω = log IAP/Ks; where Ω is the saturation state of the system, IAP is the ion activity product and Ks is the thermodynamic solubility product of the relevant phase. Calcium carbonate precipitation in microbial systems typically occurs when the saturation index (with respect to calcite) is above 1. Along with this, organics also play important role in the carbonate precipitation. The organics act as crystallization inhibitors (when in solution) and prevent the nucleation of calcium carbonate even at high SI values. It follows that bacterial presence and activity are a prerequisite for the precipitation of calcium carbonate. The bacterial metabolic activity produces the necessary increase in supersaturation as to induce the heterogeneous crystallization of calcium carbonate.

Polymorphs of $CaCO_3$.

Besides its scientific interest, calcium carbonate polymorph selection can have important technical implications. For instance, the bacterial conservation of building materials requires the formation of a coherent, durable, calcium carbonate cement i.e., calcite. This is not fully achieved if metastable vaterite (more soluble than calcite) is formed during a bacterial conservation treatment.

Applications of Biominerals

The use of bacterially induced carbonate biominerals is becoming increasingly popular day by day. From removal of heavy metals and radio nucleotides, removal of calcium from wastewater and biodegradation of pollutants, atmospheric CO_2 sequestration, remediation of building materials, modifying the properties of soil and filler in rubber and plastics to fluorescent markers in stationery ink, bacterial carbonates are serving many fields.

Removal of Heavy Metals and Radionuclide

Heavy metals and radionuclides pose greatest challenges for their disposal. They are threatening not only to human health but to the entire environment. Remediation of ground water from radionuclides has been the topic of immediate attention and several measures have being taken for it. Traditional methods of using pump and treat have been found ineffective and paved way for applications of ureolytic bacteria capable of $CaCO_3$ precipitation. In this, biomineralization of radionuclide and contaminant metals into calcite occurs as competitive co-precipitation reaction in which suitable divalent cations are incorporated into the calcite lattice.

$$^{90}Sr^{2+} + OH^- + HCO_3^- = SrCO_3(s) + H_2O$$

These cations and radionuclides integrate into the calcite structure via substitution of calcium ions in the microenvironment of the mineral precipitate, forming low - strontium carbonate minerals which have very low solubility. Researchers reported 95% capturing of the total strontium in solid phase by Microbially induced calcite precipitation (MICP) with application of *S. pasteurii*. Researchers successfully investigated the potential of enriching native ureolytic organism *in situ* for remediation of ground water by co-precipitation of Strontium.

Metal contamination of soil is especially problematic because of strong adsorption of many metals to the surfaces of soil particles. For removal of toxic metals, the conventional methodologies include phytoremediation, on site chemical leaching of contaminants, bioremediation by toxic metal tolerant bacterial species but all these treatments are expensive and not successful for long terms. Other treatments include addition of cement or chemical fixatives, by capping with asphalt, or by *in situ* vitrification and excavation and confinement of soils in hazardous waste facilities. Although rapid in effect, both of these options are expensive ($30–300/$m^3$) and destroy the soil's future productivity.

MICCP has come up as a novel solution to this problem. Recently Researchers reported the positive potential of Sporosarcina ginsengisoli CR 5 for remediation of Arsenic contaminated sites as well as copper bioremediation by Kocuria flava CR1. Researchersalso investigated the impact of varying concentrations of different metals as lead, cadmium, chromium, zinc, copper and nickel in MICP technology and reported that metal toxicity do play important role in microbial substrate degradation.

Removal of Calcium Ions and Polychlorinated Biphenyls (PCBs)

The presence of a high concentration of calcium ions (500–1500 mgL^{-1}) in the wastewater causes severe scaling in the pipelines and reactors due to calcium formation as carbonate, phosphate and/or gypsum. Researchers also studied the potential of removing Ca^{2+} ions from industrial wastewaters by MICCP. By the addition of a low concentration of urea (0–16 gL^{-1}), up to 90% of the calcium ions was successfully removed from the examined wastewater.

The last 15 years have seen an increase in the types of contaminants to which bioremediation is being applied, including polychlorinated biphenyls (PCBs). Microbial processes are beginning to be used in the cleanup of these recalcitrant contaminants. The conventional methods of removing PCBs include solvent washing, hydroblasting, or sandblasting followed by encapsulation in epoxy coating. But all these treatments are quite inconvenient and ineffective due to resurfacing of the oil over time. An alternative to these is the use of carbonate biomineralizing bacterial isolates. Researchers recently reported the positive potential of ureolytic S. pasteurii cultures and urea/calcium for treatment of PCB - coated cement cylinders leading to surficial encapsulation of PCB containing oils.

Carbon Dioxide Sequestration

The exponentially growing concentrations of carbon dioxide in the environment invited researchers from worldwide to deal with the problem. Different ways were proposed to reduce the emission of CO_2 into the atmosphere and earlier strategies developed were:

- Reduce its production and release into the atmosphere (by use of less carbon intensive energy sources like wind, solar and nuclear energy).
- Increase efficiency of energy use from production to its end use.

But both these options were not possible practically, so researchers shifted to capturing/ sequestering the CO_2 in a safe manner. Conversion of CO_2 to carbonates offers the possibility of safe, stable and environmentally benign product for long term carbon sequestration, as massive carbonate mineral reservoirs have existed for millions of years. This option was looked upon by various researchers to capture and dispose the produced CO_2 in a safe manner i.e., sequestration of CO_2. This method includes fixing the CO_2 in the form of carbonate minerals such as calcite, magnesite, and dolomite and came out to be safe and permanent method for disposing off the CO_2 as the carbonate minerals are environmentally benign and stable. The proposed mechanism for reducing emissions is the capture and storage of CO_2 in deep geologic reservoirs, such as deep saline aquifers. Leakage of CO_2 back has also to be prevented which might be due to decreased well bore integrity and increased cap rock permeability.

In nature, the sequestration of CO_2 is based on the chemical fixation of CO_2 in the form of carbonate minerals such as calcite, aragonite, dolomite and magnesite over a geological time scale. Scientists therefore proposed biomimetic CO_2 sequestration by using biological catalysts like carbonic anhydrase (CA) to minimize the localized CO_2 concentration. CA enzyme is ubiquitously distributed in organisms and fundamental to many eukaryotic biological processes like photosynthesis, respiration, CO_2 and ion transport etc. used bovine CA as enzyme to accelerate CO_2 hydration and reported that precipitation of calcium carbonate occurred much more quickly in presence of bovine CA. The role of biological CA in biological calcification of marine invertebrates (mollusks), fish otoliths, corals, hard tissues of vertebrates has been widely studied but applications of these microbes for CO_2 sequestrations have started only recently.

MICCP serves as potential solution to seal fractures and high permeability leaking zones. The storage of CO_2 is enhanced by MICCP by increasing the dissolved CO_2 (as carbonate or bicarbonate) in the subsurface formation water or the precipitation of dissolved CO_2 in carbonate minerals. Sequestration of CO_2 by bacterial CA from different bacterial isolates has been recently shown by many. Researchers also proposed MICP to protect the well cements from supercritical CO_2, plug microfractures near well environment and reduce the permeability in cap rock. Researchers also investigated S. pasteurii in artificial groundwater to determine the transformation of CO_2 into a solid carbonate phase (mineral trapping).Researchers demonstrated that as pH increases, the DIC increases and CO_2 (g) decreases. It was concluded that ureolysis-driven MICCP in the subsurface can potentially increase the security of long-term CO_2 storage. Researchers studied successful deposition of $CaCO_3$ by extracellular CAs of *M.* chthonoplastes cyanobacterial cells. Recently Researchers reported potential employment of cyanobacteria for point-source carbon capture and sequestration. The cyanobacteria utilize solar energy through photosynthesis to convert carbon dioxide to recalcitrant calcium carbonate biominerals.

Filler for Rubber, Plastics and Ink

Application of MICCP technique has been recently reported to produce a material that can be used as filler in rubber and plastics, fluorescent particles in stationery ink, and a fluorescent marker. The group isolated moderately thermophilic bacterium *Geobacillus thermoglucosidasius* from thermophilically composted organic waste. This bacterium catalyzes the formation of calcite crystals which harbor the property of fluorescence. The calcite crystals formed by *G. thermoglucosidasius* nucleation are excited by a wavelength interval of 260–400 nm, and their emission wavelengths are from 350 to 600 nm. The wide emission wavelength interval is a novel fluorescence property of *G. thermoglucosidasius* catalyzed calcite crystals which encourages it usage as filler in rubber and plastics, fluorescent particles in stationery ink, and fluorescent marker. In materials engineering, environmentally friendly systems with minimal energy consumption and resource depletion are required for producing materials and composites. Biological processes serve as impressive archetypes of sustainable materials technologies. Because of the potential benefits of biominerals in this regard, its study has gained recognition as an important area of biomimetic materials science.

Biomineralized Carbonates in Construction Materials

The potential of MICCP technology in restoration of cement mortar cubes, sand consolidation and limestone monument repair, reduction of water and chloride ion permeability in concrete, filling of pores and cracks in concrete, enhanced strength of bricks via urea hydrolysis pathway has been investigated by many researchers.

Biodeposition in Soil and Sand Materials

Mechanical properties of soil are often insufficient and subjected to erosion. Stabilization of soil is highly desirable because of the increasing infrastructure. Conventional chemical grouting techniques are often expensive and require many injection wells for treating large volumes. Recently, the techniques which aims at changing soil properties on demand by stimulating natural (bio-) chemical processes *in situ* has been found and called as biogrouting i.e., *in situ* soil strengthening technique, involving microbial induced carbonate precipitation. In this case, reagents and catalysts are injected and transported to the location where strengthening is required. Many researchers have evaluated the potential of application of bacterially induced carbonate precipitation by ureolytic bacteria by providing urea and calcium sources in various sand plugs.Researchers reported that sand consolidation by B. pasteurii reduced porosity by up to 50% and permeability by up to 90% in the areas where the cementation took place. Improvement in strength of sand columns upon bacterial carbonates was also reported. Recent research initiatives have shown that the calcite crystals form cohesive “bridges” between existing sand grains, increasing the stiffness of sand with limited decrease in permeability. Ureolytic driven MICCP has also been proposed to suppress dust, reduce permeability in granular media, improve soils, stabilize slopes and strengthen liquefiable soils. Researchers reported the successful potential of MICCP in soil for ground reinforcement. Researchers subjected the soils on shore Snake river (USA) with ureolytic biomineralization treatments and found around 1% $CaCO_3$ in near surface and 1.8–2.4% calcite below 90 cm. Researchers also found considerable reduction in permeability and improvement in shear strength of soil upon application of ureolytic bacterial sp. isolated from tropical beach sand. In the studies of many researchers halotolerant, alkaliphilic Bacillus sp. *VS1* successfully sealed a sand-lined model pond. Successive percolation treatments with high concentrations of urea and calcium solutions resulted in a nearly impermeable crust on the surface of the sand, which markedly reduced the seepage rate. Researchers recently showed the potential of MICCP by ureolytic bacteria to suppress dust.

Cemented sand body from a large scale biogrout experiment.

Biomineralization in Cementitious Materials

The successful application of bacterial biomineral calcite for improving the compressive strength of mortar and concrete has been reported by several studies. In all these works, different ureolytic bacteria have been applied and they all have improved the properties of cement concrete specimens to a considerable extent. Researchers found increase in resistance of concrete toward alkali, freeze thaw attack, drying shrinkage and reduction in permeability upon application of bacterial cells. Researchers enhanced the permeability characteristics of mortar by B. sphaericus. Researchers treated mortar cubes with *Bacillus* sp. CT-5 and reported nearly six times less absorption of water as compared to untreated specimens. They also studied the effect of B. pasteurii on water impermeability in concrete cubes and found the reduction in penetration of water which was more significant on the top side as compared to sides because of better compaction and closing of pores at the top surface.

Table: Overview of different applications where microbial calcite is used as biocement in cementitious materials.

Application	Organism	Solution
Cement mortar	Bacillus cereus Bacillus pasteurii	Nutrical SF
Remediation of cracks in concrete	Bacillus sphaericus Bacillus pasteurii Shewanella Sporosarcina pasteurii	Growth and bio cementation medium SF Sterile CSL urea medium, LML urea medium NB urea medium
Self-healing	Bacillus pseudifirmus Bacillus cophnii	Calcium lactate

Remediation of Cracks in Concrete

In case of sealing the cracks in concrete structures, Biomineralized $CaCO_3$ has proved its efficacy in many reports. Researchers used B. pasteurii for microbiologically enhanced crack remediation. Similar work was carried out by researchers who also supported microbiological remediation of cracks in concrete. Researchers encapsulated ureolytic bacterial cells in polyurethanes (PU) and reported positive potential of microbiologically enhanced crack remediation by PU immobilized bacterial cells. Researchers also reported positive potential of Biomineralized calcite in crack repair of concrete by B. sphaericus. Researchers reported that compressive strength of Biomineralized specimens could be restored to 84%. Recently Researchers reported the positive potential of bacterial calcite in crack remediation by Bacillus sp. CT-5.

Restoration of Limestone Buildings

MICCP has also been studied for its efficacy in reducing the permeation properties and thereby leading to enhancing the durability of stone specimens by various researchers.

Table: Overview of different methodologies where microbial calcite has been deposited as a layer on surface of stone.

Experimental Methods	Application Procedure		
Inoculum	Bacteria	Nutrients	Evaluation procedures
Culture in exponential phase: 10^7 to 10^9 cells/ ml	Spraying	Spraying (5 times)	Water absorption, SEM analysis, surface roughness, colorimetery and plate count
Overnight culture 10^6 cells cm^{-2}	Brushing on water saturated specimens	Wetting every day for 15 days	Water absorption, colorimeteric measurements, stone cohesion
2% Inoculums	Immersion in growing bacterial culture (shaking or stationary conditions) for 30 days		Stone cohesion, weight increase, XRD and SEM analysis, porosimetery analysis
1% Inoculums	Immersion in growing bacterial culture (intermediate wetting) for 28 days		Water absorption, SEM analysis
10^8 cells ml^{-1}	Spraying	In Carbogel	Water absorption and drying due to evaporation
n.d.*	n.d.*	Immersion in test solution or spraying (*in situ*) tests	Water absorption, colorimeteric measurements, stone cohesion, staining of newly formed calcite with Alizarin Red S and Calcein
Overnight culture 10^7 to 10^9 cells ml^{-1}	Immersion for 1 day	Immersion for 4 days	Weight increase, water absorption, gas permeability, chloride migration, carbonation, freezing and thawing, SEM and XRD analysis

It was found that bacterially induced carbonates efficiently reduced the water sorptivity of the treated stone. Researchers found that application of M. xanthus induces the precipitation of carbonates, phosphates and sulfates in a wide range of solid and liquid media Carbonate cementation by this bacterial isolate is successful up to a depth of several hundred micrometers (>500 μm) without any plugging or blocking of the pores. Researchers investigate the consolidating effect of bacterial biomineral calcite on Pietra di Lecce bioclastic limestone by use of Micrococcus sp. and B. subtilis. Researchers also reported 50% reduction in water absorption by application of B. sphaericus to limestone cubes. The use of carbogel as a delivery system for bacteria for applications in limestone specimens was proposed researchers. Researchers investigated the application of calcite crystals precipitated by fresh water bacteria on limestone and found significant reduction in pore sizes of the substrate in treated specimens. Researchers reported B. sphaericus to be very efficient strain for consolidation of limestone specimens at range of temperatures (10, 20, 28, 37°C). This isolate led to 64% lower weight loss upon sonication and 46% decreased sorptivity in treated limestone specimens compared to the control specimens. Researchers recently applied bacterial calcite in two types of stones: microporous and macroporous. They reported that application of bacterial carbonates is more successful in macroporous stone where it occurs to a larger extent and at greater depths than in microporous stone.

Rescue of Buildings of Historical Interest

Increasing environmental pollution in many areas has been endangering the survival of carbonate stones of historic importance. Architectural and sculptural stones have been seen to undergo deterioration due to several physical, chemical, and biological weathering. A classic example of art under threat was found in the cave of Lascaux in southwest France where devastating infection of Fusarium fungus and other molds covered the floor and banks of the main decorated chamber. Other paintings that shared the same fate were found in the Altamira cave in Santillana del Mar, Spain, and the earliest known Christian paintings that adorn Roman catacomb walls. To rescue these and other cultural testimonies, curators teamed up with the idea of recovery and consolidation of works of art by microbial biominerals. Although microbes were the causative agents for deterioration, they served as the solution also. Several research groups reported that the anaerobic sulphate-reducing bacteria Desulfovibrio desulfuricans and D. vulgaris efficiently removed the black sulphate crusts that often tarnished buildings. Other studies reported that the bioformation of oxalic acid generates a protective calcium oxalate patina on stone surfaces. Shewanella oneidensis MR1, also inhibited the rate of calcite dissolution under laboratory conditions.

In Low Energy Building Materials

Construction sector is responsible for major input of energy resulting in large share of CO_2 emissions into the atmosphere. The emission of these greenhouse gases during manufacturing processes of building materials is contributing a lot to global warming. It's time to put emphasis on reducing the emission of these gases into the atmosphere and save energy by minimizing usage of conventional building materials, methods, techniques and working on some other substitutes. For reduction of indirect energy use in building materials, either alternative for bricks, steel and cement have to be found, or vigorous energy conservation measures in these segments of industry have to be initiated. Energy requirements for production and processing of different building materials, CO_2 emissions and the implications on environment have been studied by many researchers.

Table: Energy in basic building materials.

Type of material	Thermal energy (MJ/ kg)
Cement	5.85
Lime	5.63
Lime Pozollana	2.33
Steel	42
Aluminium	236.8
Glass	25.8

Researchers reported soil cement blocks with 6–8% cement content to be most energy efficient building material. These materials have low cost, are easily recyclable and environmental friendly as the soils are mixed with additives like cement, lime etc. As there is no burning involved, this type of stabilized mud blocks helps in conserving huge amounts in energy. Researchers successfully improved the durability of these low energy building materials by application of ureolytic Bacillus sp. This work indicates the potential of this technology for sustainable, cheap and durable buildings.

Industrial Byproduct Building Materials

As the traditional construction materials (such as concrete, bricks, hollow blocks, solid blocks, pavement blocks and tiles) are all produced from the existing natural resources and damaging the environment due to continuous exploration and depletion of natural resources, many researchers have looked for reusing the wastes in environmentally and economically sustainable ways. Different types of wastes along with their recycling and utilization potentials are listed in Table.

Table: Different types and sources of solid wastes and their recycling and utilization potentials for construction materials.

Type of wastes	Source details	Recycling and utilization potentials
Industrial waste (inorganic)	Coal combustion residues, fly ash, steel slag, construction debris.	Bricks, blocks, tiles, cement, paint, fine and coarse aggregates, concrete, wood substitute products, ceramic products.
Agro waste (organic)	Baggage, rice and wheat straw and husk, saw mill waste, ground nut shell, jute, sisal, cotton stalk.	Cement boards, particle boards, insulation boards, wall panels, roof sheets, binder, fibrous building panels, bricks, acid proof cement, coir fiber, reinforced composites, polymer composites.

Mining/ mineral wastes	Coal washeries waste, mining waste tailing from iron, copper, zinc, gold industries.	Bricks, fine and coarse lightweight aggregates, tiles.
Non-hazardous waste	Waste gypsum, lime sludge, lime stone waste, broken glass and ceramics.	Blocks, bricks, cement clinker, hydraulic binder, fibrous gypsum boards, gypsum plaster, super sulfated cement.
Hazardous waste	Contaminated blasting materials, galvanizing waste, metallurgical residues, sludge from waste water and waste water treatment plants.	Boards, bricks, cement, ceramics, tiles.

Fly ash (FA) generated during the combustion of coal for energy production is one of the industrial byproduct that is recognized as an environmental pollutant. Rice husk ash (RHA) obtained from burning of rice husk is another major agricultural byproduct and available in all parts of the world except Antarctica. Both these materials can be used successfully in construction materials such as bricks and blocks without any degradation in the quality of products but problems associated with ash bricks are low strength, higher water adsorption, low resistance to abrasion, low fire resistance and high porosity studied the application of bacterial calcite on these ash bricks (FA and RHA) and found it to be very effective in reducing permeability and decreasing water absorption leading to enhanced durability of ash bricks.

Challenges and Issues

The field of biomineralization includes a multidisciplinary research involving experts from various fields. Though its potential has been suggested in a variety of sectors but several efforts need to be made to address key research and development questions necessary for commercial scale applications.

One of the disadvantages of microbial cementation method in comparison with chemical methods is that the microbial process is usually slower. This method is more complex than the chemical one as the microbial activity depends on many environmental factors including temperature, pH, concentrations of donors and acceptors of electrons, concentrations and diffusion rates of nutrients and metabolites etc. Due to this complexity, its usage at large-scale has not been so encouraging. Other gaps with usage of calcinogenic microbes include inconvenient application procedures, development of other unwanted microbes as repeated application of nutrients conceivably permit their development, need for detailed microbial ecology studies in order to ascertain the effects of the introduction of new bacteria etc.

The economic limitation of use of laboratory grade nutrient sources in field applications also restricts the use of this technology in several cases. Successful commercialization of the technique requires economical alternatives of the medium ingredients that cost as high as 60% of the total operating costs. Use of inexpensive materials as corn steep liquor (CSL) or lactose mother liquor (LML) may help to lower the cost of treatment.

Production of large volumes of reactants and cultures also makes this technology a bit inconvenient as compared to traditional treatments.

Production of ammonia during hydrolysis of urea poses environmental concern (toxic effects on human health, vegetation, atmospheric nitrogen deposition, leading to the eutrophication and acidification of sensitive ecosystems) as well as leads to discoloration of stone. Additional research is necessary to overcome this problem and encourage the use of MICCP for production of biominerals.

The survival of bacteria within the building material also influences the extent of calcification. Detailed microbial ecology studies are imperative in order to ascertain the effects of the introduction of new bacteria into the natural microbial communities, the development of the communities at short, mid and long-term, and the eventual secondary colonization of heterotrophic microorganisms using bacterial organic matter and dead cells, such as actinomycetes, fungi, etc.

Iron Biominerals

Minerals of iron are also known to occur in many organisms. This is due perhaps, to the important role of iron in many metabolic processes, and to the difficulty for organisms posed by the toxic products of ferrous iron oxidation by o2 together with the insolubility of ferric iron at neutral pH. Formation of iron minerals allows organisms to accumulate iron for future metabolic needs while avoiding high intracellular concentrations of ferrous iron. Other attributes of iron biominerals that are potentially useful to organisms include hardness, density and magnetism. An important class of iron biominerals is the ferric hydroxides or oxyhydroxides, which occur as amorphous, colloidal precipiates, as quasicrystalline minerals such as ferrihydrite, or as crystalline minerals such as lepidocrocite or goethite. Amorphous iron oxy-hydroxides are found in the sheaths and stalks produced by the so-called iron bacteria, Leptothrix and Galljanella, which utilize the oxidation of ferrous ions by o_2 as a source of energy.

Table: Major non-transition metal minerals formed by organisms.

Mineral	Formula	Organism/Function
Calcium carbonate:		
Calcite	$CaCo_3$	Algae/exoskeletons/Trilobites/eye lens
Aragonite	CaCo3	Fish/Gravity device/Mollusca/exoskeleton
Vaterite	CaCo3	Ascidians/spicules
Amorphous	$CaCo_3 \cdot Nh_2O$	Plants/ Ca store

Calcium phosphate: Hydroxyapatite Octa-calcium phosphate Amorphous		
	$Ca_{Lo}(Po_4)6(OH)_2$	Vertebrates/endoskeleton/ Teeth/ calcium store
	$CagH_2(Po4)_6$	Vertebrates/precursor/phase in bone Mussels/ Ca store Vertebrates/precursor phase in bone
Calcium oxalate Whewellite Weddellite		
	$CaC_2o_4 \cdot H_2O$	Plants/calcium store
	$CaC_2o_4 \cdot ZH_2O$	Plants/calcium store
Alkaline earth sulfates: Gypsum Barite Celestite		
	$CaSo_4$	Jellyfish larve/ gravity device
	$BaSo_4$	Algae/gravity device
	$SrSo_4$	Acantharia/cellular support

Amorphous oxy-hyclroxides are also found, for example, together with silica in dermal granules in the marine invertebrate Molpadia (Holothuroidea). The quasi-crystalline ferric oxy-hydroxide ferrihydrite is found in many, organisms in the cores of the iron storage protein ferritin. Ferritin proteins from several bacteria, known as the bacterioferritins, contain a somewhat less crystalline ferric oxy-hydroxide-phosphate in their cores. Highly crystalline lepidocrocite and goethite are found in the radular teeth of certain molluscs, the limpets and chitons.

Another important iron biornineral is the iron oxide magnetite, $F_{e3}o_4$. Magnetite has a cubic, inverse spinel structure and is ferrimagnetic at ambient temperature. It is also dense (5.1 g/cm3) and hard. Uniformly-sized particles of magnetite, often arranged in chains, are found in magnetotactic bacteria 12-16. The particles are often, if not always, enclosed in membrane vesicles; a particle and its enveloping membrane are known as a magnetosome. Magnetite is also found in the radular teeth of chitons, and has been found in organisms as diverse as honeybees and salmon. It is the basis for the magnetotactic response in bacteria, and its function in the radular teeth of chitons is presumably related to its hardness. The function of magnetite in organisms other than bacteria and chitons is still unknown but is possibly related to sensing of the magnitude or direction of the geomagnetic field.

The production of iron sulfides in marine sediments is a byproduct of sulfate reduction by bacteria, and the crystalline products, known as framboidal iron pyrites, have been studied for many years. intracellular, amorphous iron sulfides in some sulfate-reducing bacteria. More recently, crystalline particles of the iron sulfides greigite and pyrite, and pyrrhotite, have been reported in magnetotactic bacteria from sulfidic environments. These are related to the magnetotactic res'ponse of these organisms, and possibly to other, metabolic, functions as well. Greigite, the thiospinel of iron, is isomorphous with magnetite, and is ferrimagnetic at ambient temperature. Pyrrhotite has an hexagonal defect structure, and its magnetic properties depend on the ordering of the defects. Pyrite is nonmagnetic. Ferrous carbonate and ferrous phosphate have been detected in cultures of dissimilatory iron-reducing bacteria and magnetotactic bacteria, respectively. They are presumably byproducts of the export of ferrous ions into the medium and have no known functional significance.

Lowenstam has distinguished two modes of biomineraliztion, namely, biologically-induced mineralization (BIM) and biologically controlled mineralization (BCM) (also referred to as boundary-organized biomineralization (BOB)). In the fonner processes, cellular export of metabolic products results in extracellular mineral formation with ions in the enviroment. In the latter process, the mineral phases are deposited in or on preformed organic vesicles or matrices produced by the organism. The BIM processes are not controlled by the organism and the mineral particles typically have a large size distribution and no unique morphology. The BCM processes involve highly controlled mineralization and the particles often have a narrow size distribution. Much of the current research on biomineralization phenomena is directed at elucidating the structure of the organic macromolecules involved in the BCM processes and the nature of the organic-inorganic interactions.

The formation of ferric oxy-hydroxides by the iron-bacteria Leptothrix and Gallionella has some characteristics of BCM processes. The mineral is formed extracellularly by the hydrolysis and precipitation of ferric ions resulting from the bacterial oxidation of ferrous ions. The mineral particles themselves are indistinguishable from particles formed by purely inorganic hydrolysis. Yet the precipitate is molded into sheaths and stalks, presumably by organic molecules on the outer surface of the cell. In a strict sense, all oxidation of ferrous ions in solution by air and subsequent precipitation of ferric oxy hydroxides is a BlM process because oxygen in the atmosphere is a biological product.

Magnetite (Fe_3o_4) formation by bacteria occurs by both BIM and BCM processes, with dissimilatory iron-reducing bacteria and magnetotactic bacteria, respectively. The dissimilatory iron-reducing bacteria respire Fe^{3+} in the form of amorphous ferric oxy-hydroxide, and export ferrous ions into the environment where they presumably interact with excess ferric oxy-hydroxide to produce magnetite. The ferrous ions can also interact with carbonate in the medium to form ferrous carbonate, siderite. Magnetotactic bacteria form magnetite in intracellular membrane vesicles, from a ferric oxy-hydroxide precursor and ferrous ions. Under certain culture conditions, they can also export Fe^{2+} into the medium, resulting in the formation of ferrous phosphate, vivianite, presumably by a BIM process. The structures and magnetic properties of the magnetite particles produced by the BIM and BCM processes have characteristic differences.

Iron sulfide formation by bacteria also occurs by both ElM and BCM processes, with sulfate-reducing bacteria in marine sediments and magnetotactic bacteria from high sulfide environments, respectively. Iron biomineral formation in eukaryotic organisms is almost exclusively BCM. Ferritin formation involves a protein vesicle that controls the deposition of ferrihydrite. In this case, the protein may also function to prevent the formation of more intractable minerals such as hematite, Fe_2O_3, or magnetite, to facilitate iron mobilization from ferritin when required by the organism. Bacterioferritin functions in a similar manner. Ferritin also serves as a precursor to the formation of iron biominerals on the radular teeth of chitons and limpets. The final product, either magnetite or goethite, is deposited in an organic matrix on the tooth. Ferric oxy-hydroxide and lepidocrocite are involved at intermediate stages of the process.

Evolution of Iron Biomineralization

Isotopic analysis of Precambrian sulfide minerals shows that S, the light isotope of sulfur, is preferentially enriched compared to the heavy isotope, S, in deposits younger than 2.7 Byr BP (billion years before present). This is evidence for biological intervention in the production of sulfide, and suggests that iron sulfide biomineralization by ElM processes is at least that o1d1. Since the magnetotactic bacteria that form iron-sulfide particles are found in sulfiderich environments, BCM of iron-sulfides could be equally ancient.

BIM of iron oxides and oxy-hydroxides certainly dates back at least to 2.0 Byr BP, when free oxygen began to accumulate in the earth's atmosphere as a result of bacterial photosynthesis. Since magnetite production in bacteria can occur anaerobically as well as microaerobically, BCM of iron oxides could have originated at the same time or even earlier. In fact, reported the extraction of magnetite particles resembling those in modern magnetotactic bacteria from Precambrian rocks. Iron biominerals formed by ElM and/or BCM processes could have been incorporated into the unusual iron-rich sedimentalY deposits known as the banded-iron formations, which were deposited between 3.8 and 1.6 Byr BP.

Amorphous Silica

While silicon in form of the silicates, comes second relative to all other elements in the earth's crust with regard to quantity, it plays only a marginal role in the biosphere. This can be partially attributed to the low solubility of 'silicic acid', H_4SiO_4, and its oligomeric condensation product 'amorphous silica' ; in water between pH 1 and 9 this solubility is about 100 – 140 ppm. In the presence of cations such as calcium, aluminium or iron, the solubility decreases markedly, and in sea water, the solubility is only about 5ppm. In the biosphere, dissolved amorphous silica is resorbed by organisms and then polymerized or connected to other solid structures. Amorphous silica as biomineral is featured mainly in unicellular organisms, in siliceous (glass) sponges and in several plants, where it occurs in the cell membranes of grains, grass or horsetail in the form of 'phytolothes' with a passive deterrent function. Accumulated in deposits of diatomaceous earth, the siliceous remains of diatoms illustrate that these are a rather old group of diverse unicellular organisms. The morphology of biogenic silica is determined by membrane proteins where nucleation can take place through condensation between silicic acid or its derivatives and the hydroxyl groups of organic matrix.

Strontium and Barium Sulfates

The unicellular plankton organism Acantharia which belongs to the radiolarians features exoskeletons consisting of exactly 20 strontium sulfate single crystals which can assume very complex shapes. The origin of these forms is connected to the morphology of the cell since nonbiogenic strontium sulfate (celestite) crystallizes as flat rhombs when left undisturbed. The higher symmetry and simplicity of this special mineral have made the Acantharia a preferred object for studies of biomineralization.

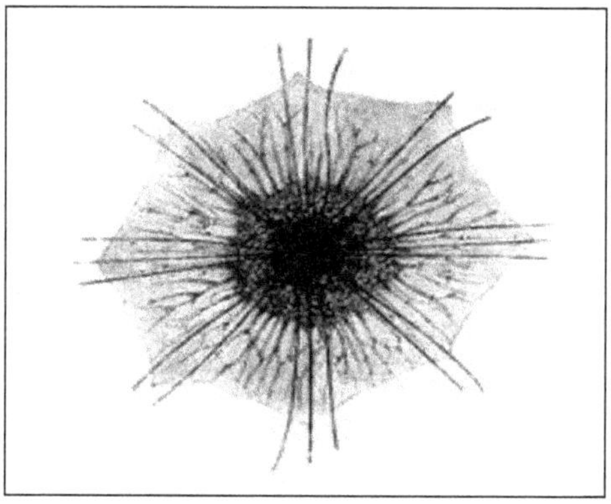

Twenty – spicule Acantharia species.

Certain unicellular algae (Desmidiacea) contain vesicles with barium sulfate in their crystalline form as baryte. The cells of Closterium, for example, are bent in a crescent-like shape with two barite crystal containing vesicles at opposing ends of the cell which suggests a function as inertia sensors. The crystals show the normal rhombic form of baryte i.e. morphology of the cell does not influence the crystal shape in contrast to the Acantharia.

References

- Biomineralization-a-confluence-of-materials-science-biophysics-proteomics-and-evolutionary-biology: cambridge.org, Retrieved 05, January 2020
- Proteins-in-calcium-phosphates-biomineralization, contemporary-topics-about-phosphorus-in-biology-and-materials: intechopen.com, Retrieved 15, June 2020

5

Physical Methods in Bioinorganic Chemistry

The physical methods which are used within the field of bioinorganic chemistry can be broadly categorized into X-ray methods, magnetic resonance methods, Mössbauer spectroscopy and vibrational spectroscopy. This chapter closely examines these physical methods as well as their usage in bioinorganic chemistry to provide an extensive understanding of the subject.

In order to understand the nature of transition metal center in biomolecules, it is important to get knowledge about its magnetic properties and electronic states. These properties are frequently correlated with the coordination capabilities and molecular geometry of the metal center, both of which affect the reactions that can occur at such a site. The magnetic and electronic phenomenon is basic need for the most spectroscopic techniques which are used to examine the chemical reactivity at metal sites in biomolecules.

X-Ray Methods

X-ray is the electromagnetic radiation of extremely short wavelength and high frequency, with wavelengths ranging from about 10^{-8} to 10^{-12} metre and corresponding frequencies from about 10^{16} to 10^{20} hertz (Hz). X-rays are commonly produced by accelerating (or decelerating) charged particles; examples include a beam of electrons striking a metal plate in an X-ray tube and a circulating beam of electrons in a synchrotron particle accelerator or storage ring. In addition, highly excited atoms can emit X-rays with discrete wavelengths characteristic of the energy level spacing in the atoms. The X-ray region of the electromagnetic spectrum falls far outside the range of visible wavelengths. However, the passage of X-rays through materials, including biological tissue, can be recorded with photographic films and other detectors. The analysis of X-ray images of the body is an extremely valuable medical diagnostic tool.

X-rays are a form of ionizing radiation—when interacting with matter, they are energetic enough to cause neutral atoms to eject electrons. Through this ionization process the energy of the X-rays is deposited in the matter. When passing through living tissue, X-rays can cause harmful biochemical changes in genes, chromosomes, and other cell components. The biological effects of ionizing radiation, which are complex and highly dependent on the length and intensity of exposure, are still under active study. X-ray radiation therapies take advantage of these effects to combat the growth of malignant tumours.

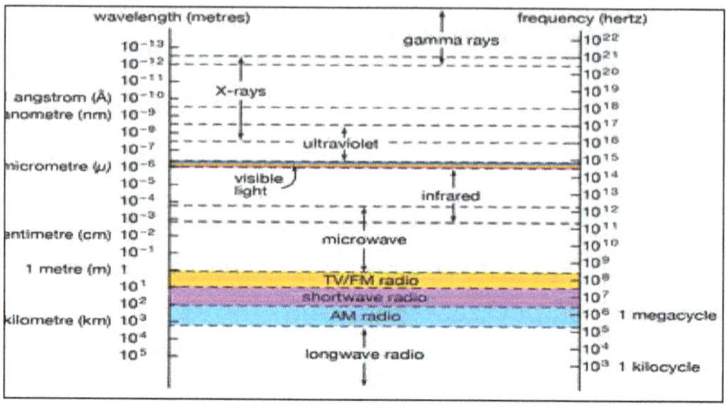

Electromagnetic spectrum: The relationship of X-rays to other electromagnetic radiation within the electromagnetic spectrum.

X-rays were discovered in 1895 by German physicist Wilhelm Konrad Röntgen while investigating the effects of electron beams (then called cathode rays) in electrical discharges through low-pressure gases. Röntgen uncovered a startling effect—namely, that a screen coated with a fluorescent material placed outside a discharge tube would glow even when it was shielded from the direct visible and ultraviolet light of the gaseous discharge. He deduced that an invisible radiation from the tube passed through the air and caused the screen to fluoresce. Röntgen was able to show that the radiation responsible for the fluorescence originated from the point where the electron beam struck the glass wall of the discharge tube. Opaque objects placed between the tube and the screen proved to be transparent to the new form of radiation; Röntgen dramatically demonstrated this by producing a photographic image of the bones of the human hand. His discovery of so-called Röntgen rays was met with worldwide scientific and popular excitement, and, along with the discoveries of radioactivity and the electron, it ushered in the study of the atomic world and the era of modern physics.

Fundamental Characteristics

Wave Nature

X-rays are a form of electromagnetic radiation; their basic physical properties are identical to those of the more familiar components of the electromagnetic spectrum—visible light, infrared radiation, and ultraviolet radiation. As with other forms of electromagnetic radiation, X-rays can be described as coupled waves of electric and magnetic fields traveling at the speed of light (about 300,000 km, or 186,000 miles, per second). Their characteristic wavelengths and frequencies can be demonstrated and measured through the interference effects that result from the overlap of two or more waves in space. X-rays also exhibit particle-like properties; they can be described as a flow of photons carrying discrete amounts of energy and momentum. This dual nature is a property of all forms of radiation and matter and is comprehensively described by the theory of quantum mechanics.

Though it was immediately suspected, following Röntgen's discovery, that X-rays were a form of electromagnetic radiation, this proved very difficult to establish. X-rays are distinguished by their very short wavelengths, typically 1,000 times shorter than the wavelengths of visible light. Because of this, and because of the practical difficulties of producing and detecting the new form of radiation, the nature of X-rays was only gradually unraveled in the early decades of the 20th century.

In 1906 the British physicist Charles Glover Barkla first demonstrated the wave nature of X-rays by showing that they can be "polarized" by scattering from a solid. Polarization refers to the orientation of the oscillations in a transverse wave; all electromagnetic waves are transverse oscillations of electric and magnetic fields. The very short wavelengths of X-rays, hinted at in early diffraction studies in which the rays were passed through narrow slits, was firmly established in 1912 by the pioneering work of the German physicist Max von Laue and his students Walter Friedrich and Paul Knipping. Laue suggested that the ordered arrangements of atoms in crystals could serve as natural three-dimensional diffraction gratings. Typical atomic spacings in crystals are approximately 1 angstrom (1×10^{-10} metre), ideal for producing diffraction effects in electromagnetic radiation of comparable wavelength. Friedrich and Knipping verified Laue's predictions by photographing diffraction patterns produced by the passage of X-rays through a crystal of zinc sulfide. These experiments demonstrated that X-rays have wavelengths of about 1 angstrom and confirmed that the atoms in crystals are arranged in regular structures.

In the following year, the British physicist William Lawrence Bragg devised a particularly simple model of the scattering of X-rays from the parallel layers of atoms in a crystal. The Bragg law shows how the angles at which X-rays are most efficiently diffracted from a crystal are related to the X-ray wavelength and the distance between the layers of atoms. Bragg's physicist father, William Henry Bragg, based his design of the first X-ray spectrometer on his son's analysis. The pair used their X-ray spectrometer in making seminal studies of both the distribution of wavelengths in X-ray beams and the crystal structures of many common solids—an achievement for which they shared the Nobel Prize for Physics in 1915.

Particle Nature

In the early 1920s, experimental studies of the scattering of X-rays from solids played a key role in establishing the particle nature of electromagnetic radiation. In 1905 German physicist Albert Einstein had proposed that electromagnetic radiation is granular, consisting of quanta (later called photons) each with an energy hf, where h is Planck's constant (about 6.6×10^{-34} joule·second) and f is the frequency of the radiation. Einstein's hypothesis was strongly supported in subsequent studies of the photoelectric effect and by the successes of Danish physicist Niels Bohr's model of the hydrogen atom and its characteristic emission and absorption spectra. Further verification came in 1922 when American physicist Arthur Compton successfully treated the scattering of X-rays from the atoms in a solid as a set of collisions between X-ray photons and the loosely bound outer electrons of the atoms.

Adapting the relation between momentum and energy for a classical electromagnetic wave to an individual photon, Compton used conservation of energy and conservation of momentum arguments to derive an expression for the wavelength shift of scattered X-rays as a function of their scattering angle. In the so-called Compton Effect, a colliding photon transfers some of its energy and momentum to an electron, which recoils. The scattered photon must thus have less energy and momentum than the incoming photon, resulting in scattered X-rays of slightly lower frequency and longer wavelength. Compton's careful measurements of this small effect, coupled with his successful theoretical treatment (independently derived by the Dutch scientist Peter Debye), provided convincing evidence for the existence of photons. The approximate wavelength range of the X-ray portion of the electromagnetic spectrum, 10^{-8} to 10^{-12} metre, corresponds to a range of photon energies from about 100 eV (electron volts) to 1 MeV (million electron volts).

Production and Detection of X-Rays

Production of X-Rays

There are three common mechanisms for the production of X-rays: the acceleration of a charged particle, atomic transitions between discrete energy levels, and the radioactive decay of some atomic nuclei. Each mechanism leads to a characteristic spectrum of X-ray radiation.

In the theory of classical electromagnetism, accelerating electric charges emit electromagnetic waves. In the most common terrestrial source of X-rays, the X-ray tube, a beam of high-energy electrons impinges on a solid target. As the fast-moving electrons in the beam interact with the electrons and nuclei of the target atoms, they are repeatedly deflected and slowed. During this abrupt deceleration, the beam electrons emit bremsstrahlung (German: "braking radiation")—a continuous spectrum of electromagnetic radiation with a peak intensity in the X-ray region. Most of the energy radiated in an X-ray tube is contained in this continuous spectrum. Far more powerful (and far larger) sources of a continuum of X-rays are synchrotron particle accelerators and storage rings. In a synchrotron, charged particles (usually electrons or positrons) are accelerated to very high energies (typically billions of electron volts) and then confined to a closed orbit by strong magnets. When the charged particles are deflected by the magnetic fields (and hence accelerated via the change in their direction of motion), they emit so-called synchrotron radiation—a continuum whose intensity and frequency distribution are determined by the strength of the magnetic fields and the energy of the circulating particles. Specially designed synchrotron light sources are used worldwide for X-ray studies of materials.

In an X-ray tube, in addition to the continuous spectrum of radiation emitted by the decelerating electrons, there is also a spectrum of discrete X-ray emission lines that is characteristic of the target material. This "characteristic radiation" results from the excitation of the target atoms by collisions with the fast-moving electrons. Most commonly, a collision first causes a tightly bound inner-shell electron to be ejected from the atom; a loosely bound outer-shell electron then falls into the inner shell to fill the vacancy. In the process, a single photon is emitted by the atom with energy equal to the difference between the inner-shell and outer-shell vacancy states. This energy difference usually corresponds to photon wavelengths in the X-ray region of the spectrum. Characteristic X-ray radiation can also be produced from a target material when it is exposed to a primary X-ray beam. In this case, the primary X-ray photons initiate the sequence of electron transitions that result in the emission of secondary X-ray photons.

In 1913 the English physicist Henry Moseley discovered a simple relationship between the wavelengths of the X-ray emission lines from a target and the atomic number of the target element—the wavelengths are inversely proportional to the square of the atomic number. Known as Moseley's law, this relationship proved to be a definitive tool in the determination of atomic numbers in the early days of atomic physics. X-ray fluorescence techniques, in which the wavelengths of characteristic X-rays are recorded following the excitation of a target, are now commonly used to identify the elemental constituents of materials.

X-ray emission is sometimes a by-product of a nuclear transformation. In the process of electron capture, an inner-shell atomic electron is captured by the atomic nucleus, initiating the transformation of a nuclear proton into a neutron and lowering the atomic number by one unit. The vacant inner-shell orbit is then quickly filled by an outer-shell electron, producing a characteristic X-ray photon. The relaxation of an excited nucleus to a lower-energy state also sometimes results in the emission of an X-ray photon. However, the photons emitted in most nuclear transitions of this type are of even higher energy than X-rays—they fall into the gamma-ray region of the electromagnetic spectrum.

Many astronomical sources of X-rays have been discovered over the past 50 years; collectively they are a rich resource of information about the universe. X-rays are emitted by the Sun's hot corona (outer atmosphere) and by the coronas of other ordinary stars in the Milky Way Galaxy. Many binary star systems emit copious X-rays; the strongest such sources produce, in the X-ray region alone, more than 1,000 times the entire energy output of the Sun. Supernova remnants are also strong sources of X-rays, which are sometimes associated with synchrotron radiation produced by high-energy charged particles circulating in intense magnetic fields and sometimes with atomic emissions from extremely hot gases (in the range of 10 million kelvins). Powerful extragalactic sources of X-rays, including active galaxies, quasars, and galactic clusters, are currently under intense scientific scrutiny; in some cases the exact mechanisms of X-ray production are still uncertain or unknown. As the Earth's atmosphere strongly absorbs X-rays, astronomical observations in the X-ray region must be made from orbiting satellites. The launch of the Chandra X-Ray Observatory in 1999 greatly advanced the observational capabilities of X-ray astronomy.

Detection of X-Rays

Photographic film was used by Röntgen as one of the first X-ray detectors, and this simple technique remains in wide use in medical applications. The process of exposure is initiated by X-ray photons ionizing radiation-sensitive silver halide crystals in an emulsion on the film surface; the resulting photochemical change of the affected crystals darkens the exposed area.

Photographic techniques, while much improved upon since the time of Röntgen and still extremely useful for qualitative applications, are not well-suited for more quantitative measurements of X-ray intensities and spectral content. A number of more effective detection methods have been developed. In a Geiger-Müller tube, or Geiger counter, incoming X-ray photons ionize atoms in a gas-filled volume. An applied high voltage induces further ionizations from collisions between liberated electrons and neutral atoms, creating an avalanche of charged particles and a large electrical pulse that is easily detected. More sophisticated detection schemes based on the ionization of gas atoms can discriminate between X-rays of different energies. Other common detection schemes rely on the ability of X-rays to produce visible fluorescence in crystals and charge separation in semiconductors.

Two fundamentally different sorts of methods using X-rays are currently widely used: diffraction and absorption.

X-Ray Diffraction

X-ray crystallography offers the most powerful structural probe of macromolecular structure including metal coordination geometries. Small molecule structures can often be determined with uncertainties in the bond distances of the order of 0.01-0.001Å. The order of precision of macromolecular structures is usually less than that for small inorganic complexes. Because of high water content, protein crystals do not diffract X-rays as small crystals do. They diffract at around 50% which is usually highly disordered. Resolution, defined by Bragg's law, whereθ is half the scattering angle between the incident and diffracted X-ray beams.

$$d = \lambda/(2\sin\theta) = 1.5418\text{Å}/(2\sin\theta)\ \text{(for Cu radiation)}$$

In order to assign the electron density to specific amino acid residues it is important to know the amino-acid sequence for metalloproteins. This requirement, in addition to the need for heavy atom derivatives to determine the phase angles for the thousands of measured reflections from which the maps are generated (isomorphous replacement technique), makes the method very laborious. Structure of nucleic acids can also be determined by X-ray diffraction methods, although it is difficult to obtain suitably diffracting nucleic acid single crystals. These problems not withdrawing, the final results are usually well worth the efforts, having afforded the best structural information currently available about metalloproteins and, to a limited extent, metal bound to nucleic acids. For the study of polycrystalline fibre of DNA, the X-ray diffraction technique is used and its results were also had great significance in the discovery of the double helical structure for DNA. For the study of this type, electron dense metal reagents have proven to be valuable. Fused-ring organic heterocycles such as ethidium bromide bind to duplex DNA by intercalation, occupying every other interbase pair site at saturation. Direct proof of this "neighbor exclusion" binding model was provided by the X-ray diffraction patterns of polycrystalline DNA fibres containing intercalated [Pt (terpy) (HET)].

X-Ray Absorption Spectroscopy

X-ray absorption spectroscopy (XAS) is the measurement of transitions from core electronic states of the metal to the excited electronic states (LUMO) and the continuum; the former is known as X-ray absorption near-edge structure (XANES), and the latter as extended X-ray absorption fine structure (EXAFS) which studies the fine structure in the absorption at energies greater than the threshold for electron release. These two methods give complementary structural information, the XANES spectra reporting electronic structure and symmetry of the metal site, and the EXAFS reporting numbers, types, and distances to ligands and neighboring atoms from the absorbing element.

X-ray absorption spectroscopy (XAS) allows us to study the local structure of the element of interest without interference from absorption by the protein matrix, water or air. Yet, X-ray spectroscopy of metallo-enzymes has been a challenge due to the small relative concentration of the element of interest in the sample. In the PS II, for example, Mn may be at the level of 10 parts per million or less. In such a case, the use of X-ray fluorescence for the detection of the absorption spectra, instead of using the transmission detection mode, has been the standard approach. The development of intense third generation synchrotron radiation X-ray sources has also permitted the study of dilute samples.

Metal complexes, as models with known structures, have been essential in order to understand the XAS of metallo-proteins. These complexes provide a basis for evaluating the influence of the coordination environment (coordination charge) on the absorption edge energy. Study of structurally well-characterized model complexes also provides a benchmark for understanding the EXAFS from metal systems of unknown structure.

The significant advantage of XAS over the X-ray crystallography is that the local structural information around the element of interest can be obtained even from disordered samples, such as powders and solution. However, ordered samples, such as membranes and single crystals, often increases the information obtained from XAS. For oriented single crystals or ordered membranes, the interatomic vector orientations can be deduced from dichroism measurements. These techniques are especially useful for determining the structures of multi-nuclear metal clusters, such as the Mn_4Ca cluster associated with water oxidation in the photosynthetic oxygen-evolving complex (OEC). Moreover, quite small changes in geometry/structure associated with transitions between the intermediate states, known as the S-states, in the cycle of the water-oxidation reaction can be readily detected using XAS. Another useful approach has been to collect complementary EXAFS measurements, for example, at both the Mn and Ca K-edges for the OEC cluster, or following a Sr → Ca replacement measuring data at the Mn and Sr K-edges. Such measurements greatly improve the information that can be obtained for multi-nuclear metal clusters, such as the Mn_4Ca cluster in PS II, as the precision of the fits can be improved by such complementary data.

X-ray absorption spectroscopy (XAS) theory has been developed to an extent that it can be applied to complicated molecules of known structure. Although it is less straightforward to apply it to the OEC, where its molecular environment is not yet precisely defined, the basic XAS equation allows us to interpret EXAFS spectra to considerable advantage. X-ray spectral properties to be expected from specified cluster geometries can be calculated and compared with experimental measurements. Density-functional theory (DFT) can be applied to issues like the stability of a proposed cluster arrangement or the likelihood of postulated reaction paths. Moreover, the time-dependent DFT calculations provide an important insight into the electronic structure of the metal site combined with the analysis of the XANES pre-edge region.

Theory of XAS

X-ray absorption spectra of any material be it atomic or molecular in nature are characterized by sharp increases in absorption at specific X-ray photon energies, which are characteristic of the absorbing element. These sudden increases in absorption are called absorption edges, and correspond to the energy required to eject a core electron into the LUMO or to the continuum thus producing a photoelectron. The absorption discontinuity is known as the K-edge, when the photoelectron originates from a 1s core level, and an L-edge when the ionization is from a 2s or 2p electron. Figure below shows a typical energy level diagram. L-edge spectroscopy is, in general, more sensitive to the electronic, structural, and the spin state changes of the metal cluster compared to the K-edge spectroscopy, however, there are experimental difficulties in applying this technique to biological samples.

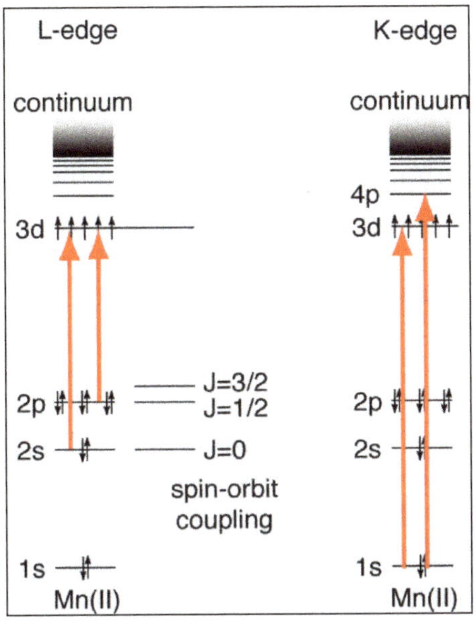

The energy level diagram for L-edge (L_I, L_{II}, and L_{III}) transitions (2s and 2p to 3d) and K-edge transitions (1s to 3d and 4p) for Mn(II). The energy levels are not drawn to scale. For example, the K-edge is at 6,539 eV and the L edges are at 769, 650, and 639 eV, respectively.

XANES

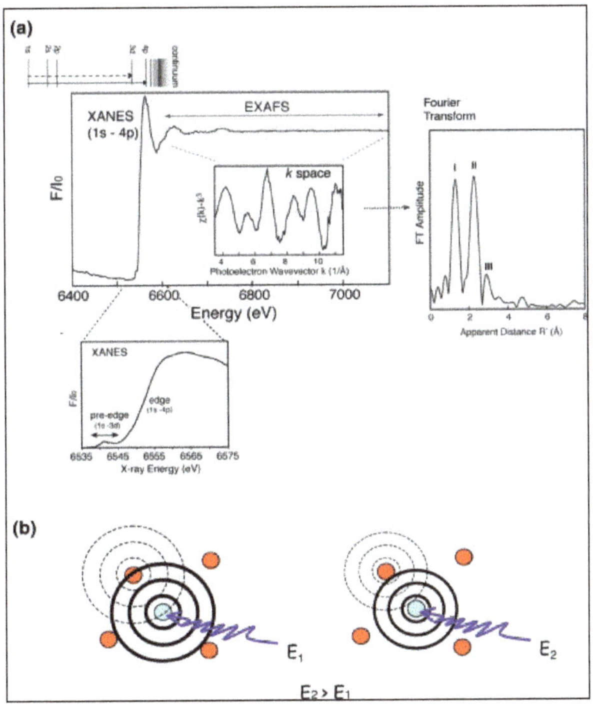

X-ray absorption near-edge structure (XANES) spectra provide detailed information about the oxidation state and coordination environment of the metal atoms. The K-edge absorption edge energy increases with increasing oxidation state. In general, the rising edge position shifts when the effective number of positive charges (in a simplified view, oxidation state) changes resulting from 1s core hole shielding effects. In an atom with one electron, for example, the electron experiences the full charge of the positive nucleus. However, in an atom with many electrons, the outer electrons are simultaneously attracted to the positive nucleus and repelled by the negatively charged electrons. The higher the oxidation state of the metal, the more positive the overall charge of the atom and therefore more energy is required to excite an electron from an orbital. Conversely, the XANES spectrum shifts to a lower energy when there is more negative charge on the metal.

In figure above, (a) The Mn K-edge XANES and EXAFS spectra. Top left: The X-ray absorption spectrum from a PS II sample showing the XANES and EXAFS regions of the spectrum. The energy levels are indicated on top of the panel. The enlargements show the Mn K-edge XANES and the k-space EXAFS spectrum. The Fourier transform of the k-space EXAFS data is shown on the right. (b) A schematic of the outgoing and backscattered photoelectron wave, which illustrates the concept of interference in EXAFS. Left: E_1 is the energy of the incident X-ray photon. The central atom (blue) is the absorbing atom and the photoelectron is backscattered from the surrounding atoms (red). The backscattered wave from the surrounding atoms (dashed blue circular lines) is in phase with the outgoing wave (solid blue circular lines). This leads to an increase in the absorption coefficient at E_1. Right: similarly, at energy $E_2 > E_1$ (notice that the wavelength of the photo-electron is shorter at E_2 compared to E_1), the backscattered wave can destructively interfere with the outgoing wave, which leads to a decrease in the cross section. The attenuation in the cross section in the absorption coefficient, called EXAFS, is a consequence of this phenomenon.

The dominant contribution to the K-edge spectrum comes from 1s → np transitions, where np represents the lowest unoccupied p orbital of the absorbing atom. This transition, with $\Delta l = 1$ (l is the orbital momentum quantum number), is quantum mechanically allowed and is typically intense. For transition metals with partially occupied d orbitals, additional insights can be gained by examination of pre-edge features that result from 1s to (n − 1)d transitions. These are relatively weak in intensity ($\Delta l = 2$; hence, formally forbidden or dipole-forbidden), but they can be detected as they occur at energies slightly less than that of the main absorption edge. The pre-edge peak intensity increases when the ligand environment is perturbed from octahedral symmetry.

EXAFS

At energies somewhat greater than the LUMO level, the absorption of an X-ray provides sufficient energy to cause the absorbing atom to release the electron (ionize). Any excess energy is carried off as translational kinetic energy, which is alternatively reflected in the wavelength associated with the electron treated as a wave phenomenon. The EXAFS modulations are a direct consequence of the wave nature of the photoelectron with the velocity v imparted to the photoelectron by the energy of the absorbed X-ray photon, which is in excess of the binding or threshold energy for the electron. The kinetic energy of the photoelectron is given by the following relation:

$$(E-E_o)=\frac{1}{2}m_e v^2,$$

where E is the X-ray photon energy, E_o is the ionization or threshold energy for the electron, and m_e is the electron mass. The EXAFS modulations are better expressed as a function of the photoelectron wave vector k(k = 2π/ λ, where λ is the wavelength given by the de Broglie relation, $\lambda = h/m_e v$, h is Planck's constant), which is expressed as follows:

$$k=\frac{2\pi}{h}\left[2m_e(E-E_o)\right]^{1/2}=0.512(E-E_o)^{1/2}$$

where E and E_o are expressed in electron volts (eV) and k has the units of inverse angstroms ($Å^{-1}$).

The wave nature of the departing electron results in interference owing to scattering off nearby atoms. Thus, the EXAFS oscillations result from the interference between the outgoing photoelectron wave and components of backscattered wave from neighboring atoms in the molecule, which start immediately past an absorption edge and extending to about 1 keV above the edge. The general definition for the EXAFS phenomenon χ(k), which is the oscillatory portion of the absorption coefficient, is the difference between the observed absorption coefficient μ(k) and the free-atom absorption coefficient $\mu_o(k)$, normalized by the free-atom contribution:

$$\chi(k)=\frac{\mu(k)-\mu_o(k)}{\mu_o(k)}.$$

One can envision the EXAFS phenomena by the help of a schematic of the outgoing and backscattered waves as shown in Figure b. As the energy of the photoelectron changes, so does the wavelength of the photoelectron. At a particular energy E_1, the outgoing and the backscattered waves are in phase and constructively interfere, thus increasing the probability of X-ray absorption or, in other words, increase the absorption coefficient. At a different energy E_2, the outgoing and backscattered waves are out-of-phase and destructively interfere, decreasing the absorption coefficient. This modulation of the absorption coefficient by the backscattered wave from neighboring atoms is essentially the basic phenomenon of EXAFS. And, Fourier transform (FT) of the modulation provides distance information describing the vector(s) between the absorbing atom and atoms to which it is bound—typically within a range limit of 4–5 Å.

A quantitative EXAFS modulation $\chi(k)$ can be expressed as follows:

$$\chi(k)=\sum_j \frac{N_j\left|f_j\left(\pi,k,R_{aj}\right)\right|}{kR_{aj}^2}\sin\left[2kR_{aj}+\alpha_{aj}(k)\right]$$

where N_j is the number of equivalent backscattering atoms j at a distance R_{aj} from the absorbing atom, $f_j(\pi, k)$ is the backscattering amplitude which is a function of the atomic number of the backscattering element j, and $\alpha_{aj}(k)$ includes the phase shift from the central atom absorber as well as the backscattering element j. The phase shift occurs due to the presence of atomic potentials that the photoelectron experiences as it traverses the potential of the absorber atom, the potential of the backscattering atom, and then back through the potential of the absorber atom.

In real systems, there is an inherent static disorder due to a distribution of distances R_{aj}, and dynamic disorder due to thermal vibrations of the absorbing and scattering atoms. Above equation is modified to include this disorder term or the Debye–Waller factor $e^{-2\sigma_{aj}^2k^2}$, where σ_{aj} is the root-mean-square deviation to give the following equation:

$$\chi(k)=\sum_j \frac{N_j\left|f_j\left(\pi,k,R_{aj}\right)\right|}{kR_{aj}^2}e^{-2\sigma_{aj}^2k^2}\sin\left[2kR_{aj}+\alpha_{aj}(k)\right].$$

The loss of photoelectrons to inelastic scattering processes can be accounted for by including a term, $e^{-2R_{aj}/\lambda_j(k)}$, which reflects losses due to inelastic scattering, where $\lambda_j(k)$ is the electron mean free path. Thus, the EXAFS contribution from each backscattering atom j is a damped sine wave in k-space, with an amplitude, and a phase, which are both dependent on k. Additionally, S_0^2 is introduced as an amplitude reduction factor due to shake-up/shake-off processes at the central atom(s). This factor can be set for fits, on the basis of fits to model compounds. Thus, the following EXAFS equation is used to fit the experimental Fourier isolates using N, R, andσ^2 as variable parameters,

$$\chi(k)=S_0^2\sum_j \frac{N_j\left|f_j\left(\pi,k\right)\right|}{kR_{aj}^2}e^{-2\sigma_{aj}^2k^2}e^{-2R_{aj}/\lambda_j(k)}\sin\left(2kR_{aj}+\alpha_{aj}(k)\right).$$

From the phase of each sine wave $[2kR_{aj}+\alpha_{aj}(k)]$, the absorber–backscatterer distance R_{aj} can be determined if the phase shift $\alpha_{aj}(k)$ is known. The phase shift is obtained either from theoretical calculations or empirically from compounds characterized by crystallography with the specific absorber–backscatterer pair of atoms. The phase shift $\alpha_{aj}(k)$ depends on both the absorber and the scatterer atoms. As one knows the absorbing atom in an EXAFS experiment, an estimation of the phase shift can be used in identifying the scattering atom.

The amplitude function contains the Debye–Waller factor and N_j, the number of backscatterers at R_{aj}. These two parameters are highly correlated, which makes the determination of N_j difficult. The backscattering amplitude function $f_j(\pi, k)$ depends on the atomic number of the scattering atom, and scattering intensity increases with the electron density (i.e., atomic number) of the scattering atom. In principle, this can be used to identify the scattering atoms. In practice, however, the phase shift and backscattering amplitude function, both of which are dependent on the identity of the backscattering atom, can be used only to identify scattering atoms that are well separated by atomic number.

The EXAFS fit-quality is evaluated using two different parameters Φ and ε^2.

$$\Phi = \sum_{1}^{N_T} \left(\frac{1}{S_i} \right)^2 \left[\chi^{expt}(k_i) - \chi^{calc}(k_i) \right]^2,$$

where N_T is the total number of data points collected, $\chi^{expt}(k_i)$ is the experimental EXAFS amplitude at k_i, and $\chi^{calc}(k_i)$ is the theoretical EXAFS amplitude at k_i. The normalization factor s_i is given by,

$$\frac{1}{S_i} = \frac{k_i^3}{\sum_j^{N_T} k_i^3 \left| \chi^{expt}(k_j) \right|}.$$

The ε^2 error takes into account the number of variable parameters p in the fit and the number of independent data points N_{ind} is given by,

$$\varepsilon^2 = \left[\frac{N_{ind}}{N_{ind} - p} \right] N_T^{-1} \Phi,$$

which is estimated from the Nyquist sampling theorem,

$$N_{ind} = \frac{2\Delta k \Delta r}{\pi},$$

where Δk is the k-range of the data and Δr is the width of the Fourier-filtered peak in Å. ε^2 provides a gauge of whether the addition of another shell to the fit is justified.

The importance of the EXAFS technique to the biochemist or structural biologist depends directly on the fact that the EXAFS modulations contain information about the distance between the absorbing and backscattering atoms within a distance of about 5 Å, as well as the identity and number of the backscattering atoms. Essentially, EXAFS analysis is used to determine the radial distribution of atoms around a particular absorbing atom, thus providing a probe for the local structure in the vicinity of the absorbing atom; for example, the metal in the active site of an enzyme. These vector lengths (distances) can be determined to a precision of 0.02 Å and much more precisely than by conventional X-ray crystallography.

Advantages of XAS

X-ray absorption spectroscopy (XAS) is element specific, so one can focus on one element without interference from other elements present in the sample. In a protein, which has more than one metal like cytochrome oxidase (Cu and Fe), or nitrogenase (Fe and Mo), it is possible to study the structural environment of each metal atom selectively. The element specificity and the fact that it is always possible to obtain an X-ray spectrum of an element also means that one 'sees' all of the metal of interest, which is present in the sample. This makes it imperative that one is sure of the biochemical homogeneity of the sample and, if there is more than one site for the same metal, to resolve the structural parameters of the different sites.

Another important advantage of XAS is that the metal of interest is never 'silent' with respect to X-ray absorption spectra. The system could be 'silent' with respect to EPR, optical, or other spectroscopic methods, but one can always probe the metal site structure by XAS.

X-ray absorption spectroscopy (XAS) is not limited by the state of the sample, because it is sensitive only to the local metal site structure. The sample can be prepared as a powder, a solution or, as is done most often, as a frozen solution for biological samples. It is not necessary to obtain single crystals of the material to examine the local structure of the metal. However, having oriented crystals such as membranes and single crystals significantly increases the structural information obtained from the XAS method.

The more important aspect is that one can either trap intermediates in the enzymatic cycle or modify the site by the addition of inhibitors or substrate or generate other chemical modifications. Such samples can be made as frozen solutions, avoiding the problems of trying to obtain single crystals. The study by this technique of trapped intermediates and treated samples has yielded insights into the mechanism of the reaction involved, in several biological systems.

Damage to biological samples by X-rays is cause for serious concern for X-ray crystallography and XAS experiments. However, with the right precautions one can successfully perform these experiments leaving the materials largely intact. The most serious damage is produced by the reaction with free radicals and hydrated electrons that are produced in biological samples by X-rays. The diffusion of the free radicals and hydrated electrons can be minimized by the use of low temperatures. The use of a liquid He flow cryostat or liquid He cryostream, where the samples are at atmospheric pressure in a He gas atmosphere, has greatly reduced the risk of sample damage by X-rays. XAS experiments require a lower X-ray dose than X-ray crystallography, and radiation damage can be precisely monitored and controlled, thus allowing for data collection from an intact metal cluster.

Limitations of XAS

It is also important to realize the intrinsic limitations of EXAFS, beyond those of a purely experimental nature. A frequent problem is the inability to distinguish between scattering atoms with little difference in atomic number (C, N, O or S, Cl, or Mn, Fe). Care must also be exercised when deciding between atoms that are apart in Z, as frequently, it is possible to obtain equally good fits using backscattering atoms which are very different in Z (e.g., Mn or Cl), but which are at different distances from the absorbing atom. This is more acute when dealing with Fourier peaks at greater distances. In bridged multinuclear centers, it is not always possible to unequivocally assign the Fourier peaks at >3 Å.

Distances are usually the most reliably determined structural parameters from EXAFS. But the range of data that can be collected, often-times due to practical reasons like the presence of the K-edge of another metal, limits the resolution of distance determinations to between 0.1 and 0.2 Å. Also it is difficult to determine whether a Fourier peak should be fit to one distance with a relatively large disorder parameter or to two distances, each having a small disorder parameter. Careful statistical analysis, taking into consideration the degrees of freedom in the fits, should precede any such analysis. The resolution in the distance Δr can be estimated from the relation that $\Delta r \Delta k \sim 1$.

Determination of coordination numbers or number of backscatterers is fraught with difficulties. The Debye–Waller factor is strongly correlated with the coordination number and one must have recourse to other information to narrow the range that is possible from curve-fitting analysis alone. It is useful to compare the spectra from the unknown complex to some known model complexes (assuming that there is evidence that the structure resembles that of the model complex) and then use Debye–Waller parameters obtained from the model complexes in the fits. This method works reasonably well, when the structure of the system being studied is well-modeled by inorganic complexes.

X-Ray Absorption Spectroscopy Studies of Photosystem II

One of the advantages of XAS is that one can potentially study the chemical events from each element which is involved in the reaction. In the OEC, Mn, Ca, and possibly Cl are the key elements we can focus on, in order to obtain the mechanistic information during the catalytic cycle.

Mn XAS

The geometric and electronic structural changes of the OEC have been studied intensively using Mn XAS. Figure below shows the Mn K-edge spectrum of each S-state of spinach PS II after deconvolution of the spectra obtained from consecutive flash illumination into pure S-state spectra, and their second derivative spectra. Traditionally, the inflection point of the rising Mn K main edge (electron 1s to 4p transition) has been used as an indicator of the oxidation states in the field of XAS.

The edge positions for each of the S-states have been quantitated by measuring the inflection point energy (IPE), given by the zero-crossing of the second derivative.

Extensive model compound studies have shown that, when Mn is oxidized by one electron in a set of Mn model compounds with similar ligands, the IPE shifts 1–2 eV to higher energy. Clear differences in absorption edge energy attributed to Mn oxidation were seen in the $S_0 \rightarrow S_1$ and $S_1 \rightarrow S_2$ transitions in the OEC, but the absorption edges for S_2 and S_3 did not show a significant difference. These results were taken to indicate the absence of Mn oxidation during the $S_2 \rightarrow S_3$ transition, although different interpretation exists.

However, one has to be aware that the edge position cannot be simply an indicator of only the oxidation state and it is problematic to conclude oxidation state changes based only on the XANES inflection point. Due to the size of the metal 4p orbital, this orbital overlaps with p orbitals of the ligands, either through σ- or π-bonding. Consequently, XANES is sensitive not only to the oxidation state but also to the ligand environment of the metal. Additionally, no definite theory is available for calculating main K-edge spectra for transition-metal complexes, owing to several factors that affect the metal p-density. Information about metal–atom oxidation state changes that is less influenced by the nature of coordinating ligands is obtained from the XES.

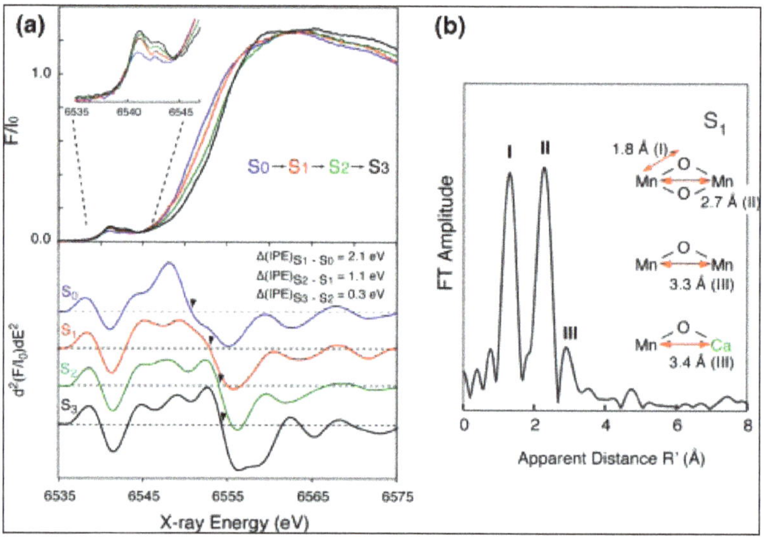

In figure above, (a) The Mn K-edge spectra of spinach PS II (BBY), from the S_0 through S_3 states (top) and their second derivative spectra (bottom). The magnitude of the inflection point energy shift for the S_0 to S_1 (2.1 eV) and S_1 and S_2 (1.1 eV) is much larger than the shift for the S_2 to S_3 transition (0.3 eV). The inset shows the pre-edge (1s to 3d transition) from the S-states is enlarged and shown above the Mn K-edge spectra. (b) The Fourier transform (FT) from a PS II sample in the S_1 state. The three FT Peak I corresponds to Mn-bridging and terminal ligand (N/O) distances at 1.8–2.0 Å, Peak II is from Mn–Mn distances (2 at ~2.7 and 1 at ~2.8 Å), and FT Peak III is from Mn–Mn distance at ~3.3 Å and Mn–Ca distances at ~3.4 Å.

The EXAFS is interpretable as shells at 1.8 and 2.0 Å (Peak I) attributable to N or O atoms and a shell at ~2.7–2.8 Å (Peak II) from Mn to Mn interactions. An additional shell from Mn was seen at 3.3 Å. The Mn EXAFS spectra changes upon the S-state transitions, particularly from the S_2 to S_3 state transition, suggesting that the OEC goes through structural changes triggered by the oxidation state changes and protonation/ deprotonation events.

Co-Factor XAS

The S-state catalytic cycle can be studied also by co-factor XAS studies. One Ca is known to be a part of the OEC, and this has been proven by Ca XAS studies and from X-ray crystallography using the anomalous diffraction technique. Regarding Cl, there is no spectroscopic evidence at least in the S_1 state that the Cl is a direct ligand to the OEC, although several biochemical studies suggest a critical role for one tightly bound Cl in maintaining oxygen-evolving activity.

In general, the requirements of X-ray spectroscopy place some restrictions with respect to sample preparation and experimental conditions. Ca and Cl in some sense fall into this category. The investigation of light elements can present difficulties due to the presence of an aqueous medium and the pervasive occurrence of C, N, and O in biological materials. In X-ray energy regions, where atmospheric gases absorb, samples must be placed in an atmosphere of helium or in vacuum. For elements like Ca and Cl, which can occur in a wide variety of environments in biological materials, it is particularly challenging to remove sources of background signals that greatly complicate interpreting the results.

Another strategy to study the role of such light element co-factor(s) is to replace it with heavier element(s). Ca can be replaced chemically or biosynthetically with Sr without losing its enzymatic activity. Similarly, Cl can be substituted with Br. XAS measurements at the Sr K-edge or Br K-edge compared to those at lower-energy Ca K-edge (~4,050 eV) or Cl K-edge (~2,850 eV) are less prone to X-ray damage, if we compare the X-ray damage by the total number of X-ray photons per area at each energy. It is also preferred for experimental reasons as X-rays at higher energies are attenuated less by the air path, the buffer solution in which the sample is made, and the cryostat windows.

Range-Extended XAS

In general, EXAFS spectra of systems which contain adjacent elements in the periodic table have a limited EXAFS range due to the presence of the rising edge of the next element, thus limiting the EXAFS distance resolution. For the Mn K-edge EXAFS studies of PS II, the absorption edge of Fe in PS II limits the EXAFS energy range. Traditional EXAFS spectra of PS II samples are collected as an excitation spectrum by electronically windowing the Kα fluorescence (2p to 1s, at 5,899 eV) from the Mn atom. The solid-state detectors that have been used over the past decade have a resolution of about 150–200 eV (FWHM) at the Mn K-edge, making it impossible to discriminate Mn fluorescence from that of Fe Kα fluorescence (at 6,404 eV). The presence of the obligatory 2–3Fe/PS II (Fe edge at 7,120 eV) limits, the data to a k-range of ~11.5 Å^{-1} (k = 0.51 $\Delta E^{1/2}$, the Mn edge is at 6,540 eV and ΔE = 580 eV). The Mn–Mn and Mn–ligand distances that can be resolved in a typical EXAFS experiment are given by,

$$\Delta R = \pi / 2k_{max},$$

where k_{max} is the maximum energy of the photoelectron of Mn.

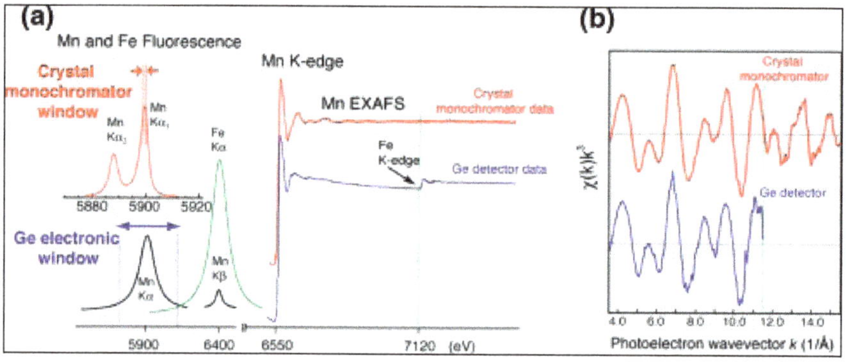

In figure, (a) Left (top): X-ray fluorescence of Mn and Fe. The multi-crystal monochromator with 1 eV resolution is tuned to the Mn Kα_1 peak (red spectrum). Left (below): Fluorescence peaks of Mn and Fe as detected using Ge-detector. The fluorescence peaks are convoluted with the electronic window resolution of 150–200 eV of the Ge-detector (black and green spectra for Mn and Fe fluorescence). Note different energy scales for the schemes shown above and below. Iron is an obligatory element in functional PS II complexes. Right: Comparison of the traditional Mn K-edge EXAFS spectrum (blue) from the S_1 state PS II sample obtained with a traditional 30-element energy-discriminating Ge-detector with a spectrum collected using the high-resolution crystal monochromator (note the absence of Fe contribution). The dashed line at k = 11.5 Å^{-1} denotes the spectral limit of a conventional EXAFS experiment owing to the iron edge. Use of the high-resolution detector eliminates the interference of Fe and removes the limit of the energy range for Mn EXAFS data collection. (b) The comparison of the k-space Mn EXAFS collected with a crystal monochromator and a Ge-detector. The range of data, as indicated by k_{max}, is inversely proportional to the resolution of the data.

The use of a high-resolution crystal monochromator allows us to selectively separate the Mn K fluorescence from that of Fe, resulting in the collection of data to higher photoelectron energies and leading to increased distance resolution of 0.1 Å. The new detection scheme produces distinct advantages: (1) improvement in the distance resolution, and (2) more precise determination in the numbers of metal–metal vectors.

These data resolve a distance heterogeneity in the short Mn–Mn distances of the S_1 and S_2 state and thereby provide firm evidence for three Mn–Mn distances between ~2.7

and ~2.8 Å. This result gives clear criteria for selecting and refining possible structures from the repertoire of proposed models based on spectroscopic and diffraction data.

Polarized XAS

Polarized XAS Studies on Oriented Membranes

Membrane proteins like PS II can be oriented on a substrate such that the lipid membrane planes are roughly parallel to the substrate surface. This imparts a one-dimensional order to these samples, while the z-axis for each membrane (collinear with the membrane normal) is roughly parallel to the substrate normal, the x and y axes remain disordered. Exploiting the plane-polarized nature of synchrotron radiation, spectra can be collected at different angles between the substrate normal and the X-ray **E** vector. The dichroism, which is the dependence of the intensity of the absorber–backscatterer pairs present in the oriented samples as a function of the polarization of the X-rays, is reflected in, and can be extracted from, the resulting X-ray absorption spectra. The EXAFS of the oriented PS II samples exhibits distinct dichroism, from which we have deduced the relative orientations of several interatomic vector directions relative to the membrane normal and derived a topological representation of the metal sites in the OEC.

To a first order approximation, the angle dependence of the EXAFS is proportional to $\cos^2(\theta_{ER})$, with θ_{ER} being the angle between the X-ray electric field vector (**E**) and the absorber–backscatter vector (**R**). In turn, θ_{ER} is composed of the detection angle θ and the angleφ between **R** and **M**, the membrane normal. Due to the rotational symmetry of the layered membranes, the angle φ defines a cone around the membrane normal **M**. When membranes are layered on a flat substrate, the preferential orientation of **M** is parallel to the underling substrate normal (**S**). For those imperfectly stacked sheets, the probability (P_α) of finding an angle α between **M** and **S** is the product of sinα and the order function $P_{ord}(\alpha)$, which is maximal at $\alpha = 0°$. $P_{ord}(\alpha)$ is approximated by a Gaussian distribution whose half-width is the mosaic spread (Ω) or the disorder angle. Here, the mosaic spread is assumed to account for the disorder between the membrane normal and substrate normal, while the spread of R relative to **M** is negligible. For an ensemble of A–B vectors (**R**), the magnitude of the EXAFS is related to the P_α- weighted integration over all possible orientations of **M** (α- and β-integration) and along the cone of the possible directions of **R** (γ-integration).

In figure, (a) Scheme for the vectors and angles involved in X-ray absorption linear dichroism of one-dimensionally oriented (PS II) membrane fragments on a flat surface. For the polarized EXAFS experiment, spectra are measured for several values of θ (angle between the X-ray electric field vector **E** and the substrate normal **S**); θ_{ER} is the angle between, **E** and the absorber–scatterer vector, **R**. θ_{ER} is composed of the detection angle θ and the angle φ between **R** and **M**, the absorber–backscatterer vector and the membrane normal. Because of the rotational symmetry of the layered membranes, the angle φ defines a cone around the membrane normal, **M**. When membranes are layered on a flat substrate, the preferential orientation of **M** is parallel to the underlying substrate normal, **S**. For an ensemble of **R** vectors, the magnitude of the EXAFS is related to the P_α-weighted integration over all possible orientations of **M** (α- and β-integration) and along the cone of possible directions of **R** (γ-integration). (b) Mn K-edge EXAFS spectra (k^3-weighted) from oriented PS II membrane samples in the S_1 state obtained with a high-resolution spectrometer (range-extended EXAFS) at orientations of 15° (green solid line) and 75° (red dashed line) of the sample normal with respect to the X-ray **E**-vector. The orientation of the X-ray **E**-vector with respect to the membrane normal is shown as an inset. (c) The structural information from the dichroism of FT peak III is illustrated showing the orientation of the average Mn–Ca vector in relation to the Mn–Mn vector. The cones represent a range for the average Mn–Ca vector(s) along the membrane normal, and the Mn–Mn vector toward the membrane plane, respectively.

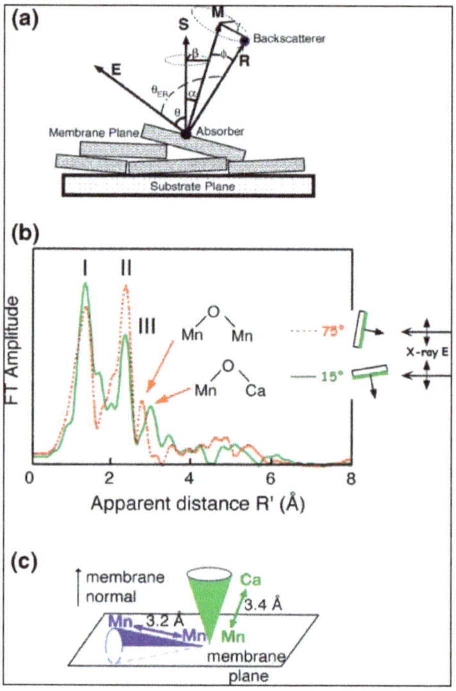

The N_{app} found from EXAFS curve-fitting on oriented samples at particular θ is related to the coordination number of an isotropic sample N_{iso} by the following equation:

$$N_{app}(\theta) = N_{iso} + \frac{1}{2} N_{iso}\left(3\cos^2\theta - 1\right)\cdot\left(3\cos^2\phi - 1\right)\cdot I_{ord},$$

where I_{ord} is the order integral:

$$I_{ord} = \frac{1}{2}\,\frac{\int_0^{\pi/2} \sin\alpha\left(3\cos^2\alpha - 1\right)\exp\left(-\alpha^2 \ln\frac{2}{\Omega^2}\right)d\alpha}{\int_0^{\pi/2} \sin\alpha \exp\left(-\alpha^2 \ln\frac{2}{\Omega^2}\right)d\alpha}.$$

By fitting the θ-dependence of N_{app} by nonlinear regression analysis, the average relative orientation φ and N_{app} can be obtained. Figure b shows the orientation of the membranes with respect to the X-ray **E**-vector and an example of the polarized spectrum from PS II.

However, as the samples are ordered in only one dimension, the dichroism information is available only in the form of an angle with respect to the membrane normal. For EXAFS measurements, this means that the absorber–backscatterer vectors can lie anywhere on a cone defined by the angle the vector forms with the membrane normal.

Polarized XAS Studies using Single Crystals of PS II

Further refinement can be performed if samples with three-dimensional order, i.e., single crystals, are examined instead of oriented membranes. Single-crystal X-ray spectroscopy has been performed on model complexes and metallo-proteins. These studies have been able to significantly expand the X-ray absorption spectroscopic information available for these systems over what is gleaned from studies of isotropic samples. An example of polarized XANES and EXAFS spectra from a Mn(V) complex is shown in Figure a and b.

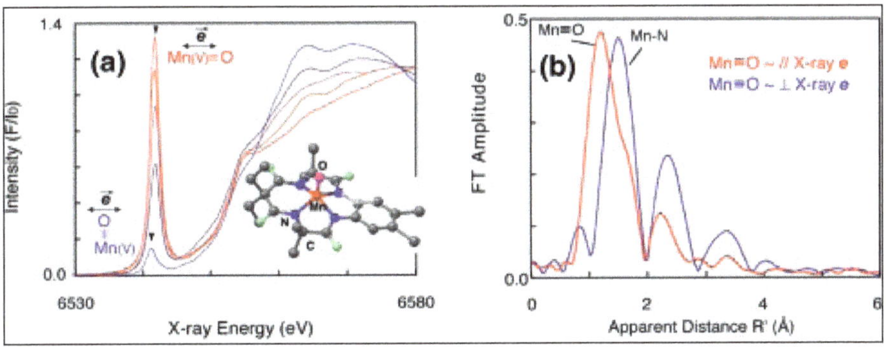

Polarized Mn XAS spectra of Mn(V)-oxo compound (inset): (a) Polarized XANES spectra. The pre-edge peak is most intense when the X-ray **E**-vector is parallel to the Mn-oxo bond. (b) Polarized EXAFS spectra in the two extreme orientations. The distinct dichroism in the XANES and EXAFS spectra show the utility of the polarized XAS methodology.

This type of analysis can also be useful for systems, where a high-resolution X-ray crystal structure is not available, such as PS II. Examination of the orientation dependence of the EXAFS of single crystals will provide structural information about the Mn sites at resolution higher than will be practically obtainable from single-crystal X-ray diffraction. Performing single-crystal EXAFS experiments can help to refine the low-resolution structure of the OEC by revealing information such as the angle(s) between the di-μ-oxo-bridged Mn–Mn vectors (~2.7 Å), as well as the relative orientation between the mono-μ-oxo Mn–Mn vector (~3.3 Å) and the di-μ-oxo-bridged Mn–Mn vectors. The directions of the Mn–Mn vectors in conjunction with the electron density derived from X-ray crystallography promises to refine the structure of the Mn complex to a resolution that neither method has presently achieved.

Figure below shows the experimental setup for collecting single-crystal XAS data from PS II at SSRL BL 9-3. It consists of a kappa goniometer, a 30-element Ge-detector for collecting XAS data, and a CCD or a MAR 345 imaging plate detector placed behind the sample for in situ collection of diffraction data to determine of the crystal orientation. The crystals are cooled using a liquid He cryostream.

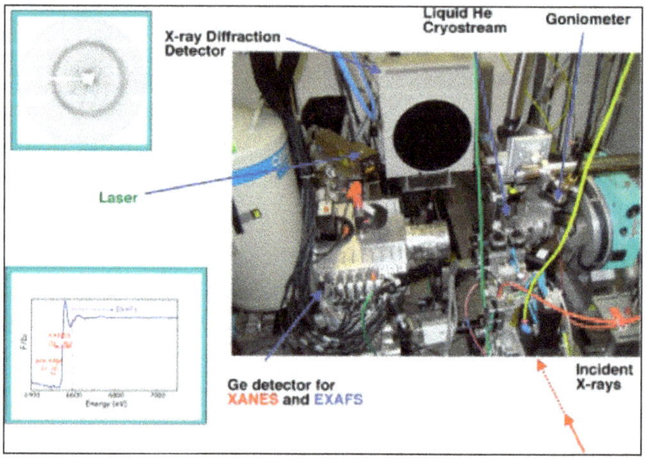

X-ray spectroscopy and diffraction set-up for PS II single crystals. The MAR345 is behind the sample, which is cooled by a liquid He cryostream to 10 K. The 30-element Ge-detector is perpendicular to the direction of the beam. The polarized EXAFS data from the single crystals of PS II improve the resolution of the distances and the determination of the directions of the vectors of the Mn complex, thus leading to a more refined structure of the Mn cluster.

Mn K-edge Pre-edge Spectra and DFT Calculations

In "X-ray absorption spectroscopy studies of photosystem II", the oxidation states of the OEC obtained from Mn XANES. In order to obtain a more detailed view of the electronic structure at the metal site, it is preferable to probe the lowest unoccupied metal 3d orbitals. The pre-edge spectra arise from excitations of 1s electron into 3d orbitals that are mainly localized around the metal ion. It shows the immediate surroundings of the excited ion through the Coulomb interaction between the core hole and the valence electrons within a short range. This pre-edge feature is a quadrupole-allowed transition; it occurs at a lower energy than the main edge transitions with approximately 1% of the intensity of the dipole-allowed main-edge transition. The transition can gain intensity by the metal 4p mixing, when the metal–ligand environment is distorted from a centro-symmetric to a non-centro-symmetric coordination. The spectra reflect coordination number, ligand environment, and oxidation state of metals. In fact, the pre-edge spectra of PS II noticeably change during the S-state transitions. In the single-crystal

XANES of PS II S_1 state, the pre-edge spectra show a characteristic dichroism. Additionally, the nature of the S_4 state can be studied by the pre-edge feature if a high-valent Mn, such as Mn(V), is involved in the transition. In order to understand the pre-edge feature and obtain the electronic configuration, however, one needs to investigate various model compounds and combine experimental data with theoretical calculations based on the ligand field and/or Density-functional theories.

Figure below shows the solution pre-edge spectrum of a five-coordinated Mn(V)-oxo model complex. Due to the strong axial distortion of the Mn site symmetry from the octahedral environment, a formally forbidden pre-edge (1s to 3d) transition gains intensity through a $3d_z$–$4p_z$ mixing mechanism and a strong pre-edge peak is observed. However, the pre-edge intensity is sensitive to the ligand environment as demonstrated in figure b by time-dependent DFT calculation of the theoretical models, in which the addition of the sixth ligand is investigated. The addition of a weak sixth ligand like water weakens the pre-edge intensity by a factor of ~2, while the addition of a stronger ligand, such as hydroxide or carboxylate, weaken the peak intensity by a factor of ~5 relative to the five-coordinated Mn(V) compounds.

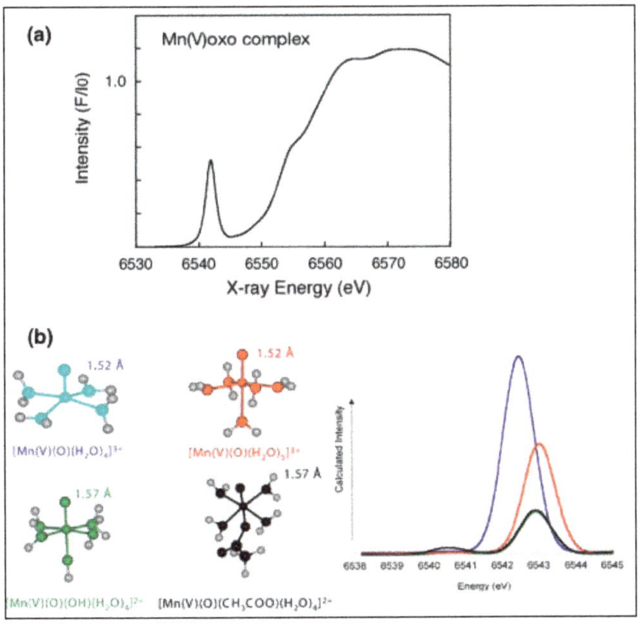

Comparison of the TD-DFT calculated Mn K-edge spectra of the Mn(V)-oxo(D-CB) complex (top) as compared to Mn(V)-oxo$(H_2O)_4$, Mn(V)-oxo$(H_2O)_5$, Mn(V)-oxo$(H_2O)_4$(OH), and Mn(V)-oxo$(H_2O)_4$$(CH_3COO)$ (bottom). The effect of the addition of a sixth ligand and/or lengthening of the Mn–O ligand distance on the intensity of the pre-edge feature is significant.

New Directions

Several unresolved problems are being addressed through new directions in XAS that are being pursued in different laboratories. Time-dependent XAS with sub-μs time resolution opens the possibility of identifying and characterizing intermediates in the individual S-state transitions that have not yet been documented. Of particular interest is the series of events on the ms time scale that accompany the formation of dioxygen during the S_4 to S_0 transition.

The combination of XAS with X-ray microscopy has shown great promise in studying very small localized domains of larger biological systems, and the possibility for combining imaging with spectroscopy. Another powerful approach has been the combined in situ use of XAS along with other methods, such as X-ray diffraction, electrochemistry, UV/Vis or FTIR/Raman spectroscopy. This methodology has allowed for monitoring of changes in the system and also the integrity of the sample. These methodologies are being applied to substrate binding studies and for following the course of catalytic reactions.

Magnetic Resonance Methods

Magnetic resonance is a phenomenon when a particle with a magnetic moment is placed in a magnetic field, give the states that have the moment oriented in different directions with respect to the applied field have different energies. Transitions between these states may be induced by absorption of light of the appropriate frequency. The appropriate light frequencies are in the range of microwave frequencies. When particles are electrons, using readily achievable magnetic field strengths, the method is called electron paramagnetic resonance spectroscopy and in the case of nuclei, the most relevant frequency range is radio-frequency region and the method is known as magnetic resonance spectroscopy.

Electron Paramagnetic Resonance (EPR) Spectroscopy

This method is ideally suited for studying many metalloproteins, including those containing Cu(II), Co(II), Fe(II), Fe(III), Mn(II), Mn(III), Mo(V), and metal clusters such as $\{Fe_2O\}^{3-}$ and $\{Fe_4S_4\}^{+,3+}$ and requires samples with unpaired electrons. The method can detect high spin ferric ions at concentrations as low as the μM range and it is also very delicate technique and it was used in the isolation and purification of iron-containing proteins, such as ferredoxin.

Purity of electron-transfer proteins cannot be established by the classic biochemical approach of enzymatic activity measurements as they do not have measurable catalytic activity. Instead, the appearance and integrated signal intensity of their EPR spectra were monitored following the usual column-chromatographic purification steps:

1. For the study of iron and copper in bioinorganic chemistry, EPR techniques are used. For, both Fe(II) and Fe(III), iron forms redox states that can exist in high and low

spin forms. EPR spectroscopy readily differentiate these configurations for ferric ion, the g and A parameters provides extra knowledge about the coordinated ligands. Vigorous EPR spectral studies of heme iron and iron-sulfur clusters have been carried out. EPR spectra have also proved to be invaluable in studying Mn(II) (I=5/2) giving a six-line hyperfine pattern), polymanganese clusters involved in photosynthesis, Mo in nitrogenase and oxotransferases, Co in vitamin B-12 and related complexes, and, most recently, Ni in the hydrogenase component of methane-generating bacteria as well as other enzymes. In addition, EPR signals are frequently encountered in organic residues adjacent to metal centers in proteins, for example the stable tyrosyl radical in ribonucleotide reductase, and in substrates such as the superoxide ion. Radical intermediates generated during enzymatic reactions can also be observed on occasion by their EPR spectra.

2. Two significant variants of the EPR method have just originated important applications in bioinorganic chemistry. One of these is electron-nuclear double resonance spectroscopy (ENDOR), in which the EPR transition is saturated while a second irradiation signal is scanned across the NMR frequency range of an isotope in the complex that might cause hyperfine splitting of the EPR lines. Usually the latter cannot be directly determined. Irradiation of the sample at definite NMR frequencies causes the EPR signal to turn into unsaturated and an ENDOR signal results. A typical application is study of ^{17}O-labeled substrates, such as H_2O, O_2, or H_2O_2, with EPR active metal centers, such as iron porphyrins.

The second method called electron-spin-echo envelope modulation (ESEEM) spectroscopy, also facilitates identification of ligand hyperfine interactions, which are sometimes called super hyperfine to differentiate them from hyperfine coupling with the metal nucleus that bears the unpaired spin. Here a series of two or sometimes three microwave pulses is applied to the sample in the EPR cavity, and the effect of varying the time delay between pulses on the "spin-echo" signal so generated is monitored. The principles are similar to those involved in NMR spin-echo experiments.

This imidazole N-H proton of a histidine ligand coordinated to Cu(II) or Fe(III), where observing the effects due to deuteration provides a perfect confirmation of the spectral assignment. Another application is to identify ^{14}N and ^{15}N nuclear super hyperfine interactions. From the magnitude of super hyperfine coupling constants, the level of electron delocalization from the metal center out onto coordinated ligands can be estimated. This information may be valuable in deducing where sites of reactivity are on bound substrate molecules.

Nuclear Magnetic Resonance (NMR) Spectroscopy

Nuclear magnetic resonance spectroscopic methods have been very beneficial in bioinorganic chemistry, as they have in other branches of chemistry, for many years. With the development of Fourier-transform spectrometers that occupy two or more time variables (two or higher dimensional NMR), the relevance of NMR spectroscopy, especially for macromolecular structural studies, has improved still further. From ^{1}H and, to a minor level, ^{13}C, ^{15}N and ^{31}P NMR spectra can be obtained distance (from nuclear Overhauser studies) and torsion angular (from coupling constants) information about the 3-D structure of biopolymers, including bound metal ions. Indeed, NMR methods can also help to obtain the complete 3-D structures of small to medium sized proteins.

One emerging area of bioinorganic chemistry that has been strongly influenced by these techniques comprises of zinc-binding domains from proteins that interact with nucleic acids. The structures of six distinct classes of such domains have now been decided by NMR spectroscopy. In all cases, the structures were determined before or contemporaneously with structural analyses of similar units by X-ray diffraction.

Paramagnetic metal ions, having electron-spin relaxation times that are fast enough not to broaden adjoining proton resonances beyond recognition, can yields shifted NMR signals that facilitate recognition of amino acid side chain ligands. This method is especially beneficial for naturally occurring Fe, Co and Cu centers. Line broadening without chemical shift changes occurs for ligands surrounding Mn(II) and Gd(III) and can be used to abstract distance information when these metals are replaced for diamagnetic metals such as Mg^{2+}, Ca^{2+} or Zn^{2+} in metalloproteins. Line broadening by paramagnetic transition metals is also valuable for shaping the accessibility of water to the active site of a metalloenzyme.

NMR methods are often quite useful for kinetic studies. The enormously high resolution and small line widths of NMR methods, in which differences in resonance position of nuclei are often measured in fractions of a part per million (ppm), allow signals to be distinguished or line broadening to be observed for species that are interconverting on a wide variety of time scales.

Mössbauer Spectroscopy

Modern approaches aimed at studying physiological and pathological processes in living systems require applications of physical techniques that can provide information at the molecular level. Mössbauer spectroscopy was already used for investigating biological materials a few years after the discovery of Rudolf Mössbauer. The development of biological applications paved the way for further applications of Mössbauer spectroscopy in medical research. Therefore, the term "biomedical" can be considered both from a broader viewpoint, which means all possible aspects of biological research, and from a narrower viewpoint, which means any biological, biophysical, and biochemical research related to medical problems (normal and pathological physiology, environmental factors and health, medicine and pharmaceutics, etc.).

Microorganisms-Related Studies

Microorganisms in soil and aquifers are well documented to be actively involved in a range of biogeochemical processes leading to biomineralization and in a diversity of chemical transformations of soil minerals. Fe(III), largely represented by extremely poorly water-soluble oxyhydroxides of variable crystallinity, is an essential component of many soil minerals. A variety of soil bacteria were reported to couple oxidation of soil organic matter, including organic pollutants, to dissimilatory reduction of many redox-active metals, primarily Fe(III) and Mn(IV).

Besides the directly biotic reduction of Fe(III) by microbial cells, a range of secondary metabolites, produced both by soil microorganisms and by plant roots and excreted into the environment, can abiotically reduce Fe (III). Some earlier and more recent experiments using ^{57}Fe Mössbauer spectroscopy have shown that such biologically and environmentally important low molecular weight organic substances as anthranilic (o-aminobenzoic) acid, indolic compounds including tryptophan, indole-3-acetic acid (IAA, a phytohormone of the auxin series produced by many soil microorganisms), and other auxins, cysteine (e.g., adsorbed on clay), and so on, can abiotically reduce Fe(III) in weakly acidic media (under pH 4–5) even under aerobic conditions.

In the case of tryptophan, no Fe(III) reduction was observed within 15 min after mixing the reagents even at pH ~2, while the resulting Fe(II) had earlier been detected in Mössbauer spectra after longer periods. In particular, partial reduction of Fe(III) to Fe(II) was detected by Mössbauer spectroscopy in the course of synthesis of ferric complexes with tryptophan and lysine, which included the procedures of precipitation and extraction with nonaqueous solvents.

Gradual Fe(III) reduction to Fe(II) by a series of auxins (indole-3-alkanoic acids with C_1–C_4 side chains) in weakly acidic aqueous solutions was evident from Mössbauer spectroscopic measurements performed in rapidly frozen solutions after 15 min and 2 days of reaction at ambient temperature. In that case, rapid freezing allows all the ongoing processes to be ceased at a certain time point, and Mössbauer measurements in the solidified frozen matrix become feasible.

Already 15 min after mixing ^{57}Fe(III) salt solution with that of an organic acid, high-spin Fe(II) was readily detectable in the spectra of filtered and frozen solutions by its characteristic quadrupole doublet with a large value of quadrupole splitting. After 2 days under those conditions, Fe(II) dominated in all the solutions, except for indole-3-carboxylic acid (ICA) for which the reduction rate was definitely lower than that for the other indolic acids studied. In all the cases, the Fe(II) quadrupole doublets gave the following parameters: isomer shift (δ), ~1.40 mm s^{-1} (here and further δ for ^{57}Fe are given relative toα-Fe at 295 K); quadrupole splitting (ΔE_Q), 3.3 mm s^{-1} (at T = 80 K), which corresponds to ferrous hexaaqua complex. Thus, the reduced iron ions in the solutions were most probably in the hydrated form not coordinated to the acids or their oxidation products.

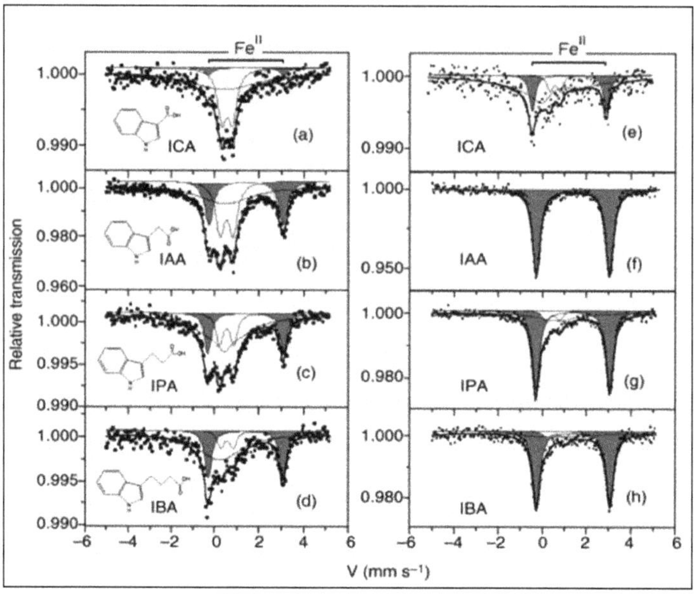

Mössbauer spectra of mixed $^{57}Fe(III)$ nitrate and indole-3-alkanoic acid aqueous solutions (molar ratios 1:3; initial $^{57}Fe(III)$ concentration 0.01 M) filtered and rapidly frozen (at T = 80 K) 15 min (a–d) and 2 days (e–h) after mixing the reagents (final pH ~3). Acids used: (a), (e)—ICA; (b), (f)—IAA; (c), (g)—indole-3-propionic acid (IPA); (d), (h)—indole-3-butyric acid (IBA).

It is noteworthy that for indole-3-acetic acid, a virtually full reduction was observed within 2 days, which corresponds to its more easily proceeding oxidative decarboxylation. However, after drying the mixture in air, all the Fe(II) was found to get reoxidized back to Fe(III). This effect, under appropriate soil conditions (acidic medium, varying humidity), could give rise to a cycling process facilitating oxidative degradation of IAA in soil in the presence of Fe(III). Since acidic soils are rather widely distributed, besides highly metal-polluted soils that often have lower pH, such abiotic Fe(III) reduction involving organic molecules of biotic origin may significantly contribute to iron transformation in soil.

Another representative ecologically important example of Fe(III) redox interaction with microbially produced extracellular biomolecules involves alkylresorcinols with short alkyl side chains—microbial signaling molecules that have been documented to perform autoregulatory and adaptogenic functions under unfavorable conditions and show other remarkably diverse biochemical and physiological activities. In aqueous mixtures of 5-methylresorcinol and $^{57}Fe(III)$ salt at final pH ~3, the appearance of Fe(II) was clearly detectable by Mössbauer spectroscopy (at 80 K) already in a few minutes, while its gradual accumulation was evident within a few hours after mixing. The parameters of the ferrous species formed upon Fe(III) reduction in the solutions (in particular, $\delta \sim 1.4$ mm s^{-1}; $\Delta E_Q \sim 3.2$ mm s^{-1} at T = 80 K) most closely correspond to hexaaquo Fe(II) coordination.

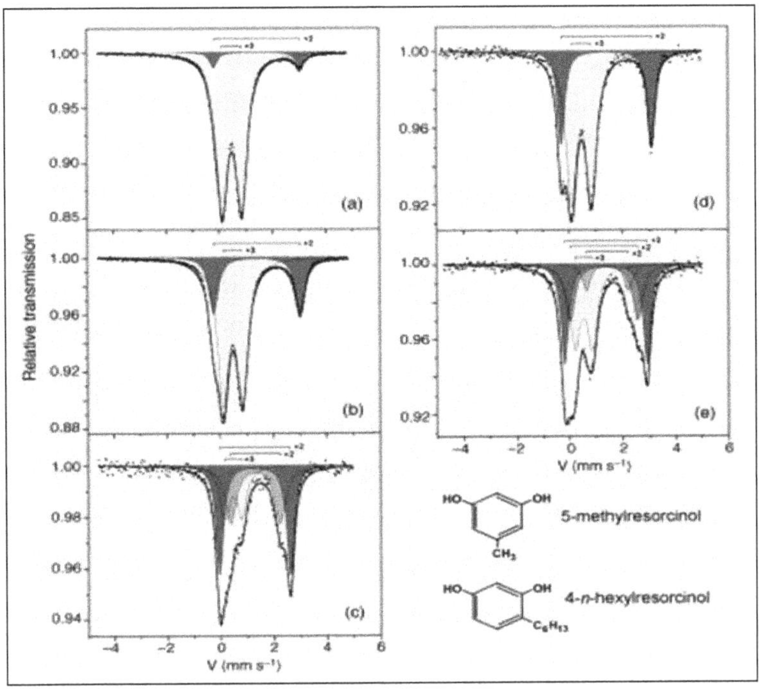

Mössbauer spectra of $^{57}Fe(III)$-containing aqueous solutions (a) of 5-methylresorcinol (orcinol) rapidly frozen 10 min and (b) 5.5 h after mixing (final pH ~3), as well as of (c) the solid residue obtained by air-drying the mixture at room temperature; (d) of $^{57}Fe(III)$-containing aqueous (20% (v/v) ethanol) solution of 4-n-hexylresorcinol rapidly frozen 5 min after mixing (final pH ~3) and (e) of the solid residue obtained by air-drying the mixture at room temperature (initial molar Fe-to-alkylresorcinol ratios 1:3; initial $^{57}Fe(III)$ concentration 0.016 M); all spectra measured at T = 80 K. The shaded areas (quadrupole doublets) represent contributions of Fe(III) (the most lightly shaded area) or Fe(II) (darker shaded areas) to the whole spectrum area (defined by the outer solid-line envelope); in spectra (c) and (e), the two and three darker-shaded areas, respectively, correspond to two different Fe(II) forms.

Drying the mixture of 5-methylresorcinol and Fe(III) in air under ambient conditions led to the resulting dried solid that gave a more complicated Mössbauer spectrum, with two different forms of Fe(II) species (altogether comprising over three-fourth of the total iron), along with a residual ferric form. The parameters of the two ferrous forms evidently represent different complexes of Fe(II) with the possible residuary 5-methylresorcinol and/or its oxidation product(s) that might include 1,2,3-trihydroxy- and/or 1,3,4-trihydroxy-5-methylbenzenes, or the relevant hydroxylated quinones. Thus, those Mössbauer measurements, in line with the UV spectroscopic monitoring of the soluble products, provide clear evidence that Fe(III) is gradually reduced by 5-methylresorcinol in moderately acidic (Ph ~3) aqueous solutions in air.

Table: Mössbauer Parameters for $^{57}Fe(III)$-Containing Aqueous Solutions of

5-Methylresorcinol and 4-n-Hexylresorcinol (Containing 20% (v/v) Ethanol) at pH ~3 (Total [Fe] = 16 ± 1 mM; 1:3 Fe-to-Alkylresorcinol Molar Ratios), Rapidly Frozen After Specified Periods of Time, and for Their Solid Residues Obtained by Drying in Air at Ambient Temperature (Measured at T = 80 K).

Alkylresorcinol Mixed with $^{57}Fe(III)$ in Solution (Sample Measured)	Time	Fe Oxidation State	$\delta^{b,c}$ (mm s^{-1})	$\Delta E_Q^{c,d}$ (mm s^{-1})	$\Gamma_{exp}^{c,e}$ (mm s^{-1})	S^f (%)
5-Methylresorcinol (solution)	10 min	+3	0.48(1)	0.74(1)	0.52(1)	93.3
		+2	1.41(1)	3.22(1)	0.34(1)	6.7
	5.5 h	+3	0.48(1)	0.72(1)	0.52(1)	74.1
		+2	1.38(1)	3.23(1)	0.42(1)	25.9
5-Methylresorcinol (dried solid)	–	+3	0.48(1)	0.63(2)	0.43(2)	24.5
		+2 (form 1)	1.31(1)	1.86(3)	0.50(3)	33.3
		+2 (form 2)	1.29(1)	2.65(1)	0.34(1)	42.2
4-n-Hexylresorcinol (solution)	5 min	+3	0.475(3)	0.757(4)	0.49(1)	68.8
		+2	1.402(3)	3.388(5)	0.34(1)	31.2
4-n-Hexylresorcinol (dried solid)	–	+3	0.54(1)	0.65(1)	0.48(2)	34.9
		+2 (form 1)	1.43(1)	1.55(3)	0.35(3)	9.0
		+2 (form 2)	1.29(1)	2.54(2)	0.45(3)	23.7
		+2 (form 3)	1.359(3)	3.10(1)	0.34(1)	32.4

Interestingly, under the same conditions, 4-n-hexylresorcinol showed a significantly higher Fe(III) reduction rate. Already 5 min after mixing the reagents, about a third of the total Fe(III) was reduced to Fe(II) represented by a quadrupole doublet with the parameters typical for hexaaquo Fe(II) complex. As Fe(III) ions at pH ~3 are well known to be prone to hydrolytic polymerization, in this case the residual Fe(III), which gives virtually the same Mössbauer parameters as those for the aqueous 5-methylresorcinol solution 10 min after mixing, represents, most probably, hydrolyzed polymeric and/or colloidal hydroxo species. Some participation of the organic molecules or their oxidation products in Fe(III) coordination cannot be ruled out merely on the basis of Fe(III) Mössbauer parameters.

After similarly drying the mixture of 4-n-hexylresorcinol and Fe(III) in air under ambient conditions, the resulting dried solid gave a yet more complicated Mössbauer spectrum, with three different forms of ferrous species (altogether comprising ca. two-third of the total iron), along with a residual ferric. The parameters of the ferrous forms in the dry residue, except probably for form 3 (which are rather close to those for hydrated Fe^{2+} ions in solid ferrous chloride hydrates at T~80 K), evidently correspond to different Fe(II) complexes with the possible residuary 4-n-hexylresorcinol and/or its oxidation product(s). However, note that these parameters and, consequently, the Fe^{2+} microenvironments (i.e., type and arrangement of donor atoms in the ligands) differ from those for the dried 5-methylresorcinol solution. Different positions of the alkyl substituents in the 1,3-dihydroxybenzene (i.e., resorcinol) moieties might be expected to lead to different oxidation products of the 5- and 4-alkylsubstituted resorcinols. This factor, possibly together with the length of the alkyl group, might be a reason for the significant difference in the Fe(III) reduction rate observed for 5-methyl-(i.e., 5-C_1-alkyl; a gradual redox process) and 4-n-hexyl-substituted resorcinol (4-n-C_6-alkyl; a much more rapid redox process). Similar trends of Fe(III) reduction in going from the autoinducer solutions to their air-dried solids have also been obtained at slightly lower pH ~1.5.

The redox behavior of Fe(III) mixtures with both alkylresorcinols upon drying was totally different from that of indole-3-acetic acid (auxin) aqueous solutions. All Fe(II) formed had been found to get reoxidized back to Fe(III). In contrast, in dried solutions of the alkylresorcinols, the resulting Fe(II) was not reoxidized upon drying in air; moreover, it was in fact found to be dominating in the dried solid. These findings point to significantly stronger reducing and/or binding (coordinating) power of alkylresorcinols or their oxidation products as compared to those of auxin.

Some microbial autoinducers can be abiotically oxidized by soil Fe(III) in moderately acidic media, which is of ecological significance with regard to microbial signaling. An oxidation process involving a particular molecular signal is very likely to exclude the latter from any signaling pathways, which is equal to "message undelivery", directly affecting microbial auto regulation and, generally, intercellular communication in the microbial consortium.

Mössbauer spectroscopy has been applied to microbial biomineralization studies in a few recent reports. For the anaerobic alkaliphilic dissimilatory Fe(III)-reducing bacterium Geoalkalibacter ferrihydriticus, reduction of amorphous ferric hydroxide (AFH), coupled to acetate oxidation, led to the formation of siderite $FeCO_3$ (at low AFH concentrations) and poorly (under [AFH] ~0.07 M) or better crystallized magnetically ordered phases (at [AFH] above 0.07 M). The analysis of Mössbauer spectra using hyperfine field distributions demonstrated that the distribution with a lower isomer shift value (δ~0.3 mm s^{-1}) could be attributed to cations Fe^{3+} located in A-positions of magnetite Fe_3O_4 and in both positions of maghemite γ-Fe_2O_3 while that with a large isomer shift value (δ~0.6 mm s^{-1}) could be attributed to cations $Fe^{2.5+}$ located in the B-positions of magnetite. It was also shown that anthraquinone-2,6-disulfonate (quinone, an additional "electron shuttle") added to the bacterial cultivation medium resulted in a more intensive AFH reduction, increasing the relative content of Fe_3O_4 and siderite. The intensity of the magnetically split sextet increased with an increase in the quinone concentration up to C_q ~ 0.05 g l^{-1}; remaining unchanged up to $C_q = 0.3$ g l^{-1}. The relaxation character of the Mössbauer spectra became more pronounced for $C_q \geq 0.4$ g l^{-1}, reflecting a reduction in the superparamagnetic particle size.

Using temperature-dependent Mössbauer measurements (at 4.5–300 K), a similar AFH reduction was reported by a binary bacterial culture including, besides G. ferrihydriticus, also the non-Fe(III)-reducing alkaliphilic bacterium Anaerobacillus alkalilacustre (in the presence of mannitol as the only substrate). The reduction process takes place due to the growth of the former bacterium that utilizes the metabolic products formed by the latter that fermented mannitol.

Fe(III) mineral characterization was reported using Mössbauer spectroscopy and X-ray diffraction for minerals formed in cell suspensions of Acidovorax sp. BoFeN_1, a nitrate-reducing, Fe(II)-oxidizing bacterium. Lepidocrocite (γ-FeOOH) (90%), which also forms after Fe(II) chemical oxidation by dissolved O_2, and goethite (α-FeOOH) (10%) were produced at pH 7.0 in the absence of strongly complexing ligands. Higher pH, increasing concentrations of carbonate or humic acids promoted goethite formation and caused little or no changes in Fe(II) oxidation rates; phosphate resulted in the formation of unidentifiable Fe(III) solids and notably slowed Fe(II) oxidation rates. Fe(III) mineralogy due to bacterial Fe(II) oxidation thus seems to be strongly influenced by the solution chemistry.

Using Mössbauer spectroscopy, three positions of trivalent iron with nonoverlapping ranges of ΔE_Q were revealed in bacterial ferrihydrite nanoparticles, produced by the bacterium Klebsiella oxytoca in the course of biomineralization of iron salt solutions from a natural medium, after sample heating. The fit of these spectra and interpretation of results seem to be doubtful, as the authors did not take into account a complicated iron core structure in bacterial ferritin and a core size variation while relating their results to merely two structural differences in ferrihydrite.

In the magnetotactic bacterium Magnetospirillum gryphiswaldense, Mössbauer spectroscopic measurements allowed a mechanism to be proposed for initial magnetite formation via membrane-bound ferritin and an Fe(II) species at the cytoplasmic membrane. The iron required for magnetite biomineralization is released directly at the membrane invagination pit interface without iron flux via the cytoplasm. This might imply distinct transport pathways for Fe_3O_4 formation and biochemical utilization of iron.

Plants

Although Mössbauer spectroscopy is capable of selectively characterizing Fe-containing chemical species, there have been relatively few Mössbauer spectroscopic studies of iron in intact plant tissues. This is mainly because of the special sample requirements for obtaining an adequate Mössbauer spectrum. The biggest constraint is a very low iron content in plants (and a low natural abundance of ^{57}Fe). In addition, for the observation of the Mössbauer Effect, the sample should be in a solid state.

Despite these limitations, the technique has been applied in the past to characterize the forms of iron occurring in different plants. More recently, a detailed study of iron uptake and translocation in rice grown in anaerobic $^{57}FeCl_2$-enriched nutrient solutions showed primarily the presence of Fe(III) hydrous oxide components precipitated on the root cell wall (ferrihydrite and/or lepidocrocite). No evidence of Fe(II) was found in the leaf tissue, the spectra were characteristic of Fe(III) present in ferritin and in other complexed forms. Iron biomineralization was also observed in a perennial grass grown in extreme acidic environment with a high content of metals. In this case, the Mössbauer spectral analysis indicated that iron accumulated in this plant mainly as jarosite and ferrihydrite (ferritin). Jarosite accumulated in roots and rhizomes, while ferritin was detected in all the structures.

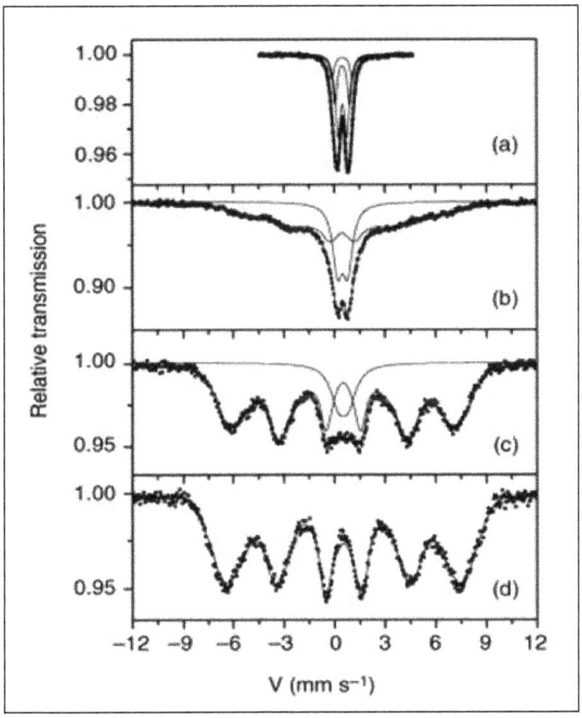

Mössbauer spectra of iron-sufficient cucumber roots grown in 10 μM ^{57}Fe(III)-citrate-containing nutrient solution recorded at (a) T = 80 K; (b) T = 4.2 K; (c) T = 1.5 K; (d) T = 1.5 K applying 5 T external magnetic field perpendicular to the propagation of the γ-ray.

Iron is an essential nutrient for plants and crucial for a variety of cell functions. In most soils, iron is not readily available for plants due to the low solubility of iron oxyhydroxides at the soil pH. For this reason, plants have evolved different strategies for iron uptake. Non graminaceous plants follow strategy I that starts with reduction of iron chelates at the outer surface of the plasma membrane. The plasma membrane-bound enzyme, ferric reductase oxidase 2, was shown to accept Fe(III) chelates and use cytoplasmic reducing equivalents ($FADH_2$[1] or NADPH[2]) to reduce them to Fe(II) chelate that releases iron due to its much lower stability, and the ferrous ions are taken up by a divalent cation transporter. The reduction rate is increased under iron deficiency and is also accompanied by an increased proton release. In contrast, graminaceous monocots can produce phytosiderophores—specific, low molecular weight Fe(III)-binding polydentate organic ligands—which solubilize iron and make it available for absorption (strategy II iron uptake mechanism).

In absorbing different species, such as Fe(III) or Fe(II), both the cell membrane and the cell wall play an important role. The complex network of cell walls within the whole plant organism is called the apoplast. It promotes the contact of cells with the environment and forms an internal space for the nonphysiological accumulation of ionic and uncharged particles. After the iron is taken up through the membranes, it enters the cytoplasm that builds a continuous network called symplast, through microscopic channels between the cells. The targets of iron assimilation are different enzymes and structures.

Roots of Fe-sufficient cucumber (Cucumis sativus) were grown in hydroponic culture in unbuffered, modified Hoagland solution containing 10 μM ^{57}Fe(III) citrate. The Mössbauer spectrum, recorded at T = 80 K, of the iron-sufficient cucumber root grown with 10 μM ^{57}Fe(III) citrate, shows an envelope of wide symmetrical quadrupole doublet that can represent several Fe(III) microenvironments. Basing on the spectral components assigned in the Mössbauer studies of plants, the spectra were fitted with three symmetrical doublets. The main component denoted by $Fe(III)_A$ represented an iron species in an octahedral coordination with $\delta = 0.50$ mm s^{-1}, $\Delta E_Q = 0.44$ mm s^{-1}, and its relative area (S) 44%. The $Fe(III)_B$ component had $\delta = 0.50$ mm s^{-1}, $\Delta E_Q = 0.80$ mm s^{-1}, and S = 35%. The third component, $Fe(III)_C$ (S = 21%), had $\delta = 0.50$ mm s^{-1}, $\Delta E_Q = 1.22$ mm s^{-1}. The Mössbauer spectrum of the freeze-dried cucumber root recorded at 4.2 K exhibited a ~70% magnetically split subspectrum, while one of the three doublet components, existing at higher temperatures, disappeared. At 1.5 K, almost all species were magnetically ordered.

The remaining paramagnetic species represented by the quadrupole doublet could be transformed into magnetically split spectrum by applying 5 T external magnetic field. The Mössbauer parameters obtained from the fit in the case of the iron-sufficient cucumber show that the major component in the roots is a high-spin Fe(III) in octahedral coordination ($Fe(III)_A$) possibly resulting from Fe(III)–carboxylate complexes. These carboxylate complexes may include Fe(III)–organic acid and Fe(III)–amino acid species inside the cell and Fe(III)–pectates in the apoplast. The Mössbauer parameters of the $Fe(III)_B$ component are in good agreement with those reported for ferrihydrite in ferritin found in several plant tissues, for example, in pea roots, or in rice. The third doublet with the highest quadrupole splitting ($Fe(III)_C$) can represent sulfate–hydroxide containing Fe(III) species. According to its characteristic Mössbauer parameters, it can be attributed to jarosite or its analogous compound ($MFe_3(OH)_6(SO_4)_2$, M = K^+, H_3O^+, Na^+,...).

However, it was not possible (not even with the help of the low-temperature measurements) to identify unequivocally which compounds are represented by $Fe(III)_B$ and $Fe(III)_C$, since the magnetic ordering temperatures strongly depend on the particle sizes and crystallinity of the Fe(III) aggregates and are in strong correlation with the stoichiometry of these compounds. The magnetic properties of hydrous ferric oxides formed via biomineralization processes (as in the case of the iron-storage protein ferritin) also strongly depend on its composition.

The high PO_4^{3-} ratio and the incorporation of organic ions such as citrate can lead to an amorphous structure and to a decrease in the antiferromagnetic ordering temperature (<4 K). Considering that the iron uptake in strategy I-type plants is accompanied by a significant proton release and that a high amount of sulfate ions is present in the nutrient solution, the formation of sulfate–hydroxide containing Fe(III) species with various stoichiometry can be favored in the cell wall. These findings support that the magnetically split subspectra at 4.2 and 1.5 K can be attributed to Fe(III) hydroxosulfates and hydrous ferric oxides, respectively. However, the interpretation of the magnetically split subspectrum for one superparamagnetic Fe-rich phase, such as ferritin with a small number of Fe atoms in the clusters, probably containing other ligands such as PO_4^{3-}, SO_4^{2-}, and organic acids, cannot be entirely excluded. The paramagnetic component, which can be transformed to magnetically split spectrum by applying 5 T external magnetic field at 1.5 K, supports the presence of ferric carboxylates, since the parameters of the quadrupole doublet (δ = 0.5 mm s^{-1}, ΔE_Q = 0.6 mm s^{-1}), at 4.2 and 1.5 K, correspond to those of the $Fe(III)_A$ component.

Although cucumber belongs to strategy I group, no Fe(II) could be detected when enough iron was available for the plant, which is in good agreement with the previous Mössbauer results. This can be explained by the high relative amount of Fe(III) present in the root tissues of iron-sufficient plants (suppressing the small contribution of the possible Fe(II) in the Mössbauer spectrum). Three main types of iron-bearing species were suggested (sulfate–hydroxide containing ferric species in the apoplast, hydrous ferric oxides possibly present as ferritin inside the cell, and Fe(III) carboxylate complexes both in the symplast and apoplast) to be formed naturally in the roots of cucumber grown in unbuffered Fe(III)–citrate nutrient solution.

Mössbauer spectra (recorded at T = 80 K) of iron-deficient cucumber roots, supplied with 500 μM ^{57}Fe(III) citrate for different time periods, as indicated in the Figure (a), and supplied with different ^{57}Fe(III) citrate concentrations, for 30 min (b).

Cucumbers were grown in iron-deficient nutrient solutions and supplied with ^{57}Fe(III) citrate at different concentrations (10, 50, 100, and 500 μM) for 30 min and at 500 μM for different time intervals (30, 60, 90, 180 min, and 24 h) only before harvest. Mössbauer spectra of the cucumber roots taken after iron supply (T = 80 K). The Mössbauer spectra of iron-deficient cucumber roots supplied with 500 μM ^{57}Fe(III) citrate for 30 min could be decomposed similar to the iron-sufficient samples, but in this case, an Fe(II) component could also be found besides the $Fe(III)_A$ and $Fe(III)_C$ components ($Fe(III)_B$ did not appear in this case). The relative amounts of the different iron species depended on the duration of iron supply: the Fe(II) showed a significant decrease with the iron treatment, while the relative amount of the total Fe(III) increased. Application of different iron concentrations during iron supply in 30-min time interval did not change the components that can be assigned to the Mössbauer spectra of iron-deficient plants supplied with 500 μM $^{57}Fe(III)$ citrate. The relative amount of the detected Fe(II) showed a correlation with the concentration of the iron present in the nutrient solution. With increasing the amount of Fe(III) citrate, the amount of the reduced iron increased as well.

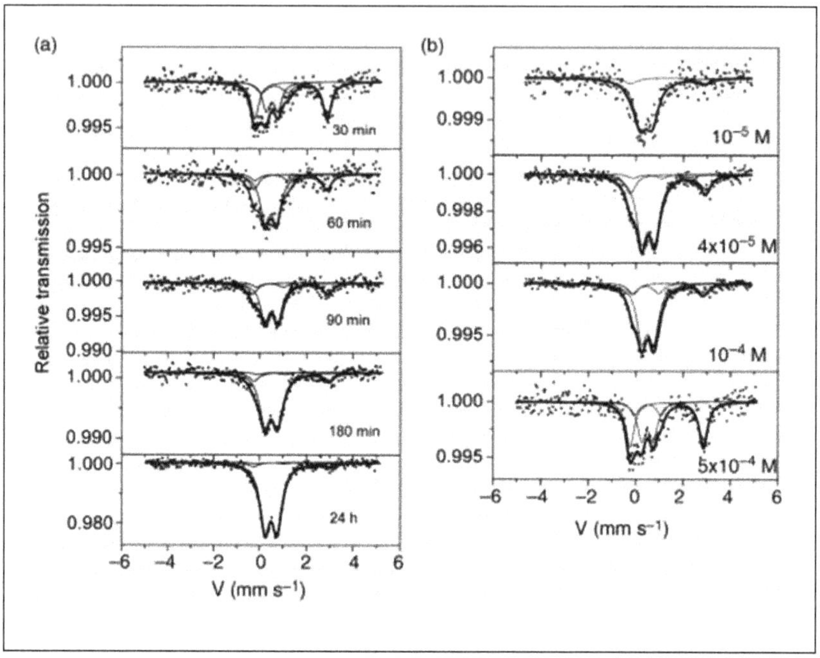

Table: Mössbauer Parameters of the Iron Components Found in the Iron Deficient Cucumber Roots that € were Supplied with 500 μM $^{57}Fe(III)$ Citrate for Different Time Intervals and with Different Amounts of $^{57}Fe(III)$ Citrate for 30 min.

	Fe(II) Component			$Fe(III)_A$ Component			$Fe(III)_C$ Component		
Sample	δ_1^e (mm s^{-1})	ΔE_{Q1}^e (mm s^{-1})	S_1^f (%)	δ_2^e (mm s^{-1})	ΔE_{Q2}^e (mm s^{-1})	S_2^f (%)	δ_3^e (mm s^{-1})	ΔE_{Q3}^e (mm s^{-1})	S_3^b (%)
Different time intervals during iron supply									
30 min	1.35	3.19	48	0.49	0.46	32	0.52	1.11	20
60 min	1.31	3.05	23	0.46	0.44	60	0.45	1.15	17
90 min	1.30	3.19	23	0.50	0.50	69	0.44	1.14	9
180 min	1.35	3.17	15	0.50	0.48	81	0.50	1.14	4
24 h	1.37	3.23	6	0.51	0.52	91	0.47	1.12	3
Different iron concentrations during iron supply									
10^{-5} M	1.31	3.22	13	0.43	0.50	87	–	–	–
5×10^{-5} M	1.35	3.15	23	0.52	0.55	72	0.50	1.23	5
10^{-4} M	1.36	2.96	16	0.51	0.51	71	0.43	1.09	13
5×10^{-4} M	1.35	3.19	48	0.49	0.46	32	0.52	1.11	20

The fact that iron(II) was clearly detected by Mössbauer spectroscopy gives a direct evidence for the iron uptake mechanism of strategy I. As the time of the iron treatment increased, the reductive capacity was decreasing because of the increased amount of iron already taken up by the root. According to the Mössbauer parameters of the Fe(II), one can see that it formed a hexaaquo complex that might be the primary hydrated product of the ferric chelate reductase enzyme, accumulated in the apoplast and not attached to any of the cell wall components. At the same time, the increase in the $Fe(III)_A$ component, representing iron both attached to the apoplast and taken up inside the cell, could be observed. The ferritin-like component (denoted by $Fe(III)_B$) was absolutely no detectable, which means this duration of iron supply was insufficient for producing the iron-storage protein. The accumulation of the iron/sulfate–hydroxide component was possibly also not favored in this case, since $Fe(III)_C$ showed a decreasing tendency.

Similar results were shown by different iron concentrations applied during iron supply but more information could be obtained about the mechanism of the reduction and the iron uptake. The relatively low amount of Fe(II) in the case of the lowest applied ^{57}Fe(III) citrate concentration can be explained by a very fast uptake and reoxidation process. If more Fe(III) was available, the Fe(II)/Fe(III) ratio was shown to increase. Consequently, the accumulation of Fe(II) in the apoplast was favored because of limited uptake and/or reoxidation processes. Comparing the total iron concentration in the root with the amount reducible in 30 min, it can be assumed that the reduction rate was high enough to potentially reduce all Fe accumulated in the apoplast. Concerning that the uptake rate of Fe(II) is limited, this might result in the accumulation of Fe(II) outside the plasma membrane.

Enzymes

Iron–sulfur clusters are versatile electron transfer cofactors, which are ubiquitous in many metalloenzymes. In the Bacillus subtilis, redox regulator (Fnr) that controls genes of the anaerobic metabolism in response to low oxygen tension, an unusual structure for the oxygen-sensing $[4Fe\text{-}4S]^{2+}$ cluster was detected by a combination of genetic experiments with UV–visible and Mössbauer spectroscopy. Asp-141 was identified as the fourth iron–sulfur cluster ligand besides three Cys residues.

In the oxygen-tolerant [NiFe] hydrogenase from the hyperthermophilic bacterium Aquifex aeolicus, three iron–sulfur clusters were identified in their usual redox states, a [3Fe-4S] and two [4Fe-4S], but also a unique high-potential state was found. On the basis of ^{57}Fe Mössbauer spectroscopy, this high-potential form was attributed to a superoxidized state of the [4Fe-4S] center proximal to the [NiFe] site. The unique environment of this cluster, characterized by a surplus cysteine coordination, is able to tune the redox potentials and make it compliant with the $[4Fe\text{-}4S]^{3+}$ state. It is actually the first example of a biological [4Fe-4S] center that physiologically switches between 3+, 2+, and 1+ oxidation states within a very small potential range.

Nonheme Fe is a conservative component of the Q-type photosynthetic reaction centers but its function remains unknown. Using Mössbauer spectroscopy, it was shown that in Rhodospirillum rubrum the nonheme Fe exists mostly in a ferrous low-spin state. The binding of Cd^{2+} ions in the vicinity of the quinone–Fe complex changed the high-spin state of the nonheme Fe into a low-spin one characterized by hyperfine parameters similar to those obtained for the nonheme Fe low-spin state in untreated reaction centers, as confirmed by Mössbauer measurements.

A fur deletion mutant of M. gryphiswaldense biomineralized fewer and slightly smaller magnetite crystals than did the wild type. Although the total cellular iron accumulation of the mutant was decreased owing to reduced magnetite biomineralization, it exhibited an increased level of free intracellular iron, which was bound mostly to a ferritin-like metabolite that was found significantly increased in Mössbauer spectra of the mutant.

The mechanism of a bacterial phenylalanine hydroxylase, a mononuclear nonheme iron protein that uses tetrahydropterin as the source of the two electrons needed to activate O_2 for the hydroxylation of phenylalanine to tyrosine. Mössbauer spectra of samples prepared by freeze-quenching the reaction of the complex enzyme–$^{57}Fe(II)$–phenylalanine–6-methyltetrahydropterin with O_2 revealed the accumulation of an intermediate at short reaction times (20–100 ms). The Mössbauer parameters of the intermediate ($\delta = 0.28$ mm s^{-1}, $\Delta E_Q = 1.26$ mm s^{-1}) suggested it to be a high-spin Fe(IV) complex similar to those that have previously been detected in the reactions of other mononuclear Fe(II) hydroxylases.

Low-temperature molecular dynamics simulations of cytochrome c oxidase predict an experimentally observable Mössbauer spectral width. Predicted lineshapes were used to model Lorentzian doublets, with which published cytochrome c oxidase Mössbauer spectra were simulated. Molecular dynamics-imposed constraints to spectral lineshapes allowed useful information to be obtained, for example, the presence of multiple chemical species in the binuclear center of cytochrome c oxidase. Molecular dynamics-based predictions were found to be in good agreement with published experimental spectra, showing promise for a renewed interest for application of this approach in bioenergetics.

Hemoglobin

Hemoglobin is the iron-containing oxygen-transport heme protein. Vertebrate hemoglobins are tetramers consisting of two pairs of nonidentical protein chains. Human and animal adult hemoglobins consist of two a-subunits and twoβ-subunits. Each subunit binds heme (iron protoporphyrin IX complex) that is the active site of hemoglobin. Heme Fe(II) reversibly binds an oxygen molecule to realize protein function. Myoglobin, an oxygen-storage heme protein, is similar to β-chains.

Owing to the important role of the heme iron electronic structure in the process of reversible oxygen binding, Mössbauer spectroscopy of hemoglobins was related to the studies of the heme iron electronic structure variations in relation to the heme iron stereochemistry. These studies were based on some features of Mössbauer spectra of hemoglobins. It was observed that Mössbauer spectra of tetrameric oxyhemoglobin had a non-Lorentzian line shape in the liquid nitrogen temperature region. This asymmetry was better fitted using a superposition of two quadrupole doublets. It was supposed that two quadrupole doublets may be a result of the heme iron electronic structure nonequivalence in α- and β-subunits of oxyhemoglobin. Mössbauer spectra of oxy-, deoxy- and carboxyhemoglobins as well as theoretical quantum chemical calculations of quadrupole splitting for the deoxyheme models in different subunits demonstrated the necessity of taking into account the heme iron electronic structure nonequivalence in α- and β-subunits of tetrameric oxy- and deoxyhemoglobins. A comparative study of human, rabbit, and pig oxyhemoglobins with different molecular structure and functions was carried out using Mössbauer spectroscopy with a high velocity resolution.

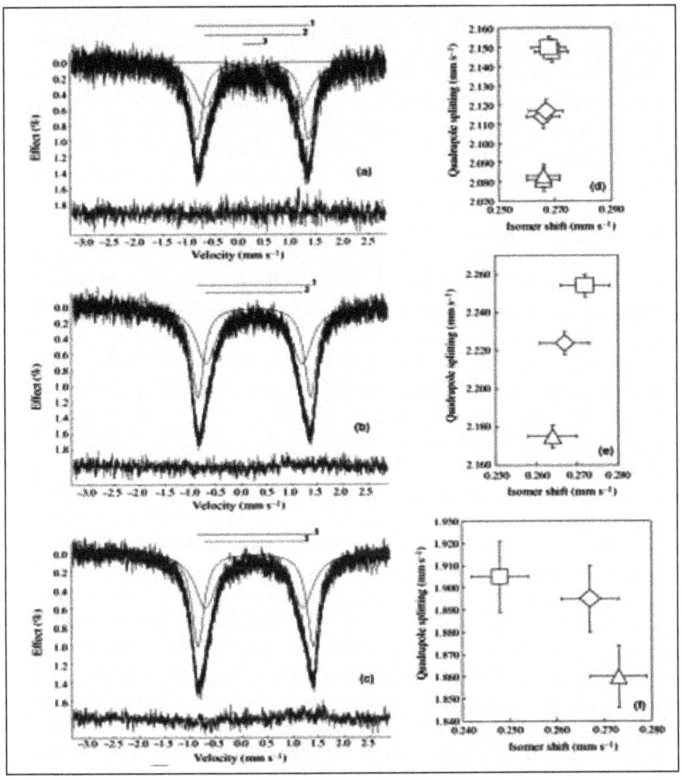

Mössbauer spectra of normal human adult (a), rabbit (b), and pig (c) oxyhemoglobins measured at 90 K in 4096 channels and presented in 1024 channels (components are the results of the best fit: 1—α-subunits, 2—β-subunits, 3—carboxyhemoglobin) and differences of Mössbauer hyperfine parameters for normal human adult (Δ), rabbit (◇), and pig ( ) oxyhemoglobins at T = 90 K (approximation with one quadrupole doublet (d) and with two quadrupole doublets: for a-subunits (e) and for b-subunits (f); indicated errors are instrumental errors for hyperfine parameters evaluated for the spectra presented in 1024 channels or calculated errors obtained during the spectra fitting if these errors exceeded instrumental errors).

Mössbauer spectra of human, rabbit, and pig oxyhemoglobins measured at 90 K and presented in 1024 channels were fitted with and without taking into account the heme iron electronic structure nonequivalence in α- and β-subunits. In both cases of spectra fits, small variations of hyperfine parameters were observed. Small differences in quadrupole splitting values of high and low oxygen-affinity human adult deoxyhemoglobin encapsulated in wet silica gel. However, this result was obtained using Mössbauer spectroscopy with a low velocity resolution (in 256 channels); therefore, the estimated error of 0.02 mm s^{-1} for quadrupole splitting (i.e., probably the error of the fit) was less than the real instrumental error which means that the obtained differences may be doubtful.

A number of Mössbauer studies were related to the analysis of the effect of external factors on hemoglobin. Some differences were shown in the processes of oxyhemoglobin radiolysis related to the type of radiation, its energy, and irradiation dose. The study of conformational changes in methemoglobin and metmyoglobin by low temperature X-ray irradiation was carried out using Mössbauer spectroscopy. Mössbauer spectra showed a reduction of high- spin Fe(III) by thermalized electrons and a production of a metastable low-spin Fe(II) myoglobin and hemoglobin complexes where H_2O was still bound at the sixth coordination.

Although the same total amount of oxidized hemoglobin was found in both cells, a well-defined oxidation product of hemoglobin was demonstrated by Mössbauer spectra only in neonatal cells. This oxidation product was not methemoglobin but a trivalent, high-spin iron compound. Further hemoglobin oxidation by acetylphenylhydrazine was used for diagnostic tests on the basis of Mössbauer spectroscopy. It was shown that red blood cells of patients affected by breast cancer are more sensitive to the denaturing action of acetylphenylhydrazine, hence the formation time of hemin, one of the last oxidation products, is significantly shorter then in normal subjects. The experiments showed differences during iron oxidation between breast cancer patients and healthy persons. Different amounts of hemin, after the same incubation time, can discriminate between samples collected from people suffering from cancer and a healthy population.

Human hemoglobin oxidation with Fe(III)-heme formation as a result of the vanadyl ion effect was observed using Mössbauer spectroscopy. Partial oxidation of rat hemoglobin after 70 days of rats feeding with cerium chloride was detected by Mössbauer spectroscopy.

Ferritin and Hemosiderin

It is well known that iron in living systems is stored in ferritin and hemosiderin. Researcher studies the process of iron core formation in human and horse spleen ferritins using Mössbauer spectroscopy. It was demonstrated that iron deposition within ferrihydrite core in human and horse spleen ferritin started with Fe(II) oxidation. This process was associated with ferroxidase center of H-chains. Further, an Fe(III) compound and Fe(III) μ-oxo-bridged dimers in ferroxidase centers of H-chains were found, which were intermediate compounds in the process of iron oxyhydroxide core formation in horse spleen ferritin. The steps leading to ferrihydrite core formation in human L- and H-ferritins were also identified and transfer between ferritin molecules was established.

To obtain information about peculiarities of the iron core structure, a number of Mössbauer investigations were undertaken using applied large magnetic fields and low temperatures. The influence of the environment on the iron core formation was investigated in bacteria grown in different solutions ($FeSO_4$, $ZnSO_4$, $TbCl_3$) and apoferritin obtained using D_2O or H_2O solution. Another Mössbauer study was undertaken to determine whether or not the iron core formation depends on the oxidant type. Researcher studied horse spleen apoferritin reconstituted at various temperatures with different numbers of iron atoms per protein shell. It was suggested that the iron cores had structures based on the mineral ferrihydrite and were superparamagnetic with a mean blocking temperature of 34 K. The hemosiderin particles had a very poor crystalline structure.

Ferritin contains the polynuclear hydrous ferric oxide core, while iron–polysaccharide complexes modeling ferritin consist of the β-FeOOH core inside the polysaccharide shell instead of the protein shell. These complexes are of particular interest because they are used or being developed as drugs for iron-deficiency anemia treatment. Among them there are iron–dextrin, iron–dextran, iron–polymaltose, and other iron–polysaccharide complexes. Some of these ferritin models were recently studied using Mössbauer spectroscopy. Mössbauer spectra of some iron– polysaccharide complexes measured by researcher demonstrated different superparamagnetic behavior and the necessity of several paramagnetic and/or magnetic components for an adequate fit of the spectra. Recently, a researcher carried out a series of comparative studies of small variations in the hyperfine parameters in human liver ferritin and various developed and commercial iron–dextran complexes (IDC) using Mössbauer spectroscopy with a high velocity resolution and spectra presentation in 512 channels. Small variations of quadrupole splitting were found at 87 K for various IDC and ferritin in frozen solutions and lyophilized forms.

Mössbauer spectra of human liver ferritin (a), Imferon (b), Maltofer (c), and Ferrum Lek (d) measured at T = 295 K with a high velocity resolution and presented in 2048 channels (a, b, d) and in 4096 channels (c), indicated components are the results of the best fit; and relationship between the spectral hyperfine parameters for Imferon ( ), Maltofer (Δ), Ferrum Lek (•), and human liver ferritin (◇) obtained using one quadrupole doublet fit (e).

To get a better insight into the issue, Mössbauer spectroscopy with a high velocity resolution was used for a comparative study of human liver ferritin and commercial medicines Imferon (iron–dextran complex), Maltofer, and Ferrum Lek (iron–polymaltose complexes). Mössbauer spectra of ferritin and its models were measured at 295 K and presented in 4096 or 2048 channels. The rough fitting of these spectra using one quadrupole split doublet (homogeneous model) demonstrated small differences of hyperfine parameters. The necessary number of quadrupole doublets (heterogeneous model) for the best fit was different for ferritin and its models. For instance, three doublets for Imferon, four doublets for ferritin, and five doublets for Maltofer and Ferrum Lek were necessary to get the best fit of the spectra measured at 295 K. These differences may be related to structural and/or size variations in the iron cores of the materials studied. The obtained results indicated that the iron core structure in human ferritin and its models is more complicated than it was believed and consists of several regions with structural variations. It was supposed that these regions can be related, for instance, to surface and several internal regions or to the presence of different crystallites.

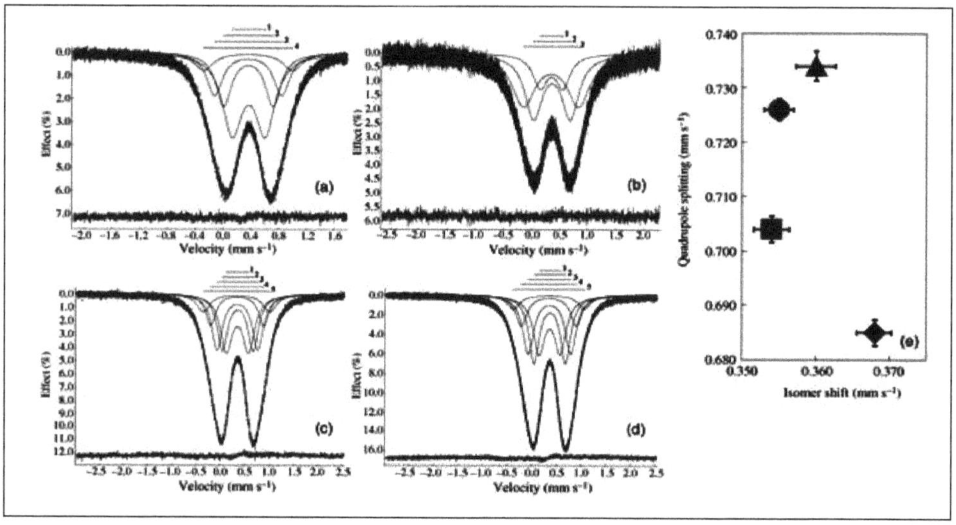

A further development in the field of Mössbauer spectra fitting and analysis is expected regarding the explanation of applicability of either continuous distributions of quadrupole splitting and hyperfine field or a superposition of discrete quadrupole doublets and magnetic sextets with models of multi-domain and multilayer structures of the ferritin iron core. In this case, application of Mössbauer spectroscopy with a high velocity resolution may be used because it leads to a lower instrumental error in the determination of hyperfine parameters (this allows small variations of hyperfine parameters to be distinguished) as well as to a more reliable fitting of complicated Mössbauer spectra).

Tissues

Mössbauer studies of tissues containing iron-storage proteins are more complex mainly because of low iron content in comparison with that in isolated ferritin and hemosiderin. In the first comparative study of purified human liver ferritin, chicken liver and spleen tissues carried out by a researcher, small differences in Mössbauer parameters were observed at 295 K. First, Mössbauer spectra were measured with a high velocity resolution in 4096 channels and presented in 512 channels. Variations of Mössbauer parameters indicated small differences in the iron electronic structure and the iron core structure of the iron-storage proteins. To increase the accuracy and decrease the instrumental error in evaluating the hyperfine parameters, Mössbauer spectra of chicken liver and spleen tissues were measured with a high velocity resolution at 295 K and presented in 1024 channels with fitting using a superposition of two quadrupole doublets within the heterogeneous model. Differences in Mössbauer parameters for chicken liver and spleen are clearly seen in Figure and indicate the presence of structural differences in the proteins' iron cores in these tissues.

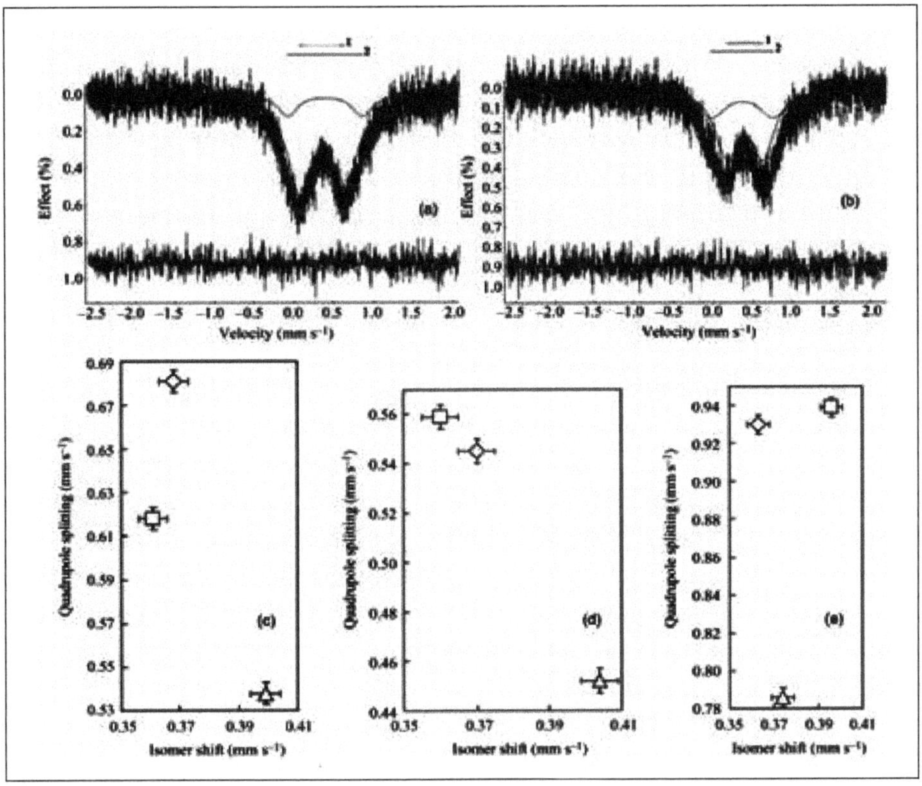

Mössbauer spectra of chicken liver (a) and spleen (b) tissues measured at 295 K and presented in 1024 channels and differences of Mössbauer hyperfine parameters for normal human liver ferritin (◇), chicken liver ( ), and spleen (Δ) tissues obtained by fitting with one (c) and two (d and e) quadrupole doublets.

The possible role of iron in Parkinson's, Alzheimer's, and progressive supranuclear palsy diseases was also studied using Mössbauer spectroscopy. Mössbauer studies demonstrated that the most of iron in the brain including three brain tissues substantia nigra, globus pallidus, and hippocampus was ferritin-like iron. The amount of ferritin molecules in Alzheimer's tissues was found to be higher than in control samples, whereas the total amount of iron was the same. It was supposed that either iron cores in Alzheimer's samples were smaller than in the control samples or a great amount of hemosiderin occurs in pathological tissues. The presence of up to 8% of nonferritin-like iron in parkinsonian substantia nigra or possible slight differences in the composition of the ferritin-like iron cores in parkinsonian substantia nigra were supposed. All Mössbauer spectra measured at 300 and 90 K looked like asymmetric doublets that were fitted using two singlets with the same area and different widths. To assess this asymmetry, the ratio of singlet widths was used referred to as the asymmetry coefficient. However, this approach has no physical meaning. Moreover, there are a number of studies in which asymmetrical Mössbauer spectra of ferritin were fitted using a superposition of several quadrupole doublets.

On the other hand, there are studies presenting an increase of iron concentration in parkinsonian tissues, so this problem remains controversial. Such iron increase was associated with neuromelanin, supposed to be the main iron-storage compound in substantia nigra neurons. It was believed that a larger cluster size demonstrated by Mössbauer spectroscopy in iron-loaded neuromelanin in comparison with iron-depleted neuromelanin and intact nigral tissue confirmed that additional iron was added to existing iron clusters in neuromelanin.

Pharmaceutical Products

Various compounds containing Mössbauer isotopes are developed and used in pharmacy. Some diorganotin(IV) compounds with cytotoxic activity against human adenocarcinoma HeLa cells were studied using ^{119}Sn Mössbauer spectroscopy by a researcher. Researchers studied organotin(IV) complexes with heterocyclic thioamides with antitumor activity using ^{119}Sn Mössbauer spectroscopy. In addition, some recent studies of tin compounds using ^{119}Sn Mössbauer spectroscopy can be mentioned. Gold compounds are also used in pharmacy. Therefore, ^{197}Au Mössbauer spectroscopy was used for the characterization of such compounds. The ^{197}Au Mössbauer spectrum of seleno-auranofin (SeAF), an analogue of auranofin (AF), the orally active antiarthritic gold drug in clinical use, exhibited an asymmetric quadrupole doublet that may be a result of a residual sample texture.

Iron-containing pharmaceutical compounds are used and developed as anticancer and antibiotic drugs as well as for treatment of iron deficiency. For instance, Mössbauer spectroscopy was used for studying iron complexes with anticancer antibiotics such as anthracyclines, adenosine derivatives, and cyclin-dependent kinase inhibitors. A Researcher analyzed iron complexes of hexachlorocyclodiphosph(V)azanes of sulfaguanidine (H_4L) with antimicrobial activity. Recently, some iron-containing complexes with antituberculosis activity were studied using Mössbauer spectroscopy.

A number of ferric and ferrous compounds are used and developed for treatment of iron-deficiency anemia. The results of Mössbauer studies of ferric complexes with various polysaccharides modeling ferritin were considered above. Recently, a characterization of the iron state in iron–polygalacturonate complex that may be used for iron supplementation was made by a researcher. One ferric and two ferrous compounds in the studied complexes were found. Various vitamins and dietary supplements contain ferrous compounds such as ferrous fumarate, ferrous sulfate, ferrous gluconate, and some others. The studies of the iron in ferrous gluconate demonstrated the presence of a minor ferric compound in addition to ferrous gluconate; moreover, it was shown that Fe^{2+} in both ferrous gluconate and Ascofer occupied two or three distinct sites. It was found that vitamins and dietary supplements with ferrous fumarate contained minor ferric and ferrous impurities, while those with ferrous sulfate contained minor ferric impurity. In contrast, vitamin and dietary supplement with ferrous iron chelates contained large amount of ferric compounds. Then these studies were continued by measuring Mössbauer spectra with a high velocity resolution. The applied experimental technique and the results obtained may be useful and important for controlling the iron state and impurity level in iron-containing medicaments.

Vibrational Spectroscopy

All molecules are constantly vibrating, and can absorb energy from an incoming photon to increase their vibrations. The two types of vibrational spectroscopy are infrared spectroscopy and Raman spectroscopy. Vibrational spectroscopy is the science of measuring exactly which wavelengths of light are absorbed by a molecule. This technique could be used identify an unknown molecule by comparing its absorption to that of other molecules. Or vibrational spectroscopy could be used to gain further understanding of the physical properties of a known molecule.

Molecular vibration can be modeled by balls attached by springs. Displacing an atom from its most stable position requires energy proportional to the displacement. IR spectroscopy can be used to characterize a molecule is the energy of its vibrations falls in the infrared range. Alternatively, higher energy light can be absorbed, the re-emitted at a different wavelength and/or in a different direction; this leads to Raman spectroscopy. So a molecule can absorb radiation to change bond lengths or positions with respect to the other atoms in the molecule.

From the number, frequency, and intensity of these absorptions or emissions we can gain insight into the composition of the sample being measured. The number of different absorptions is indicative of the number of different atoms and shape of the molecule. The frequency or wavelength absorbed is indicative of the energy of the bonds and vibrations. And the intensity of the absorptions is related to the concentration of the analyte.

Vibrational Modes

The different possible vibrations are called vibrational modes. Vibrational modes are determined by all the different ways the atoms in the molecule can move with respect to each other, called the vibrational degrees of freedom. Vibrational degrees of freedom differ from the total degrees of freedom in that translation (movement through space) and rotation do not contribute to the vibrational degrees of freedom.

To find the number of vibrational modes one must first know the point group of the molecule. From there, the point group's character table will list all of the possible symmetry operations for the molecule. Each symmetry operation will leave some atoms in the molecule in place and/or move other atoms in the molecule. Count up the number of atoms that do not move for each symmetry operation, and multiply that number by the symmetry operation's contribution. This gives the total representation of atomic motion. From there, the vibrational modes can be found by reducing the total representation, according to the equation:

$$n=1/h*\Sigma XR*XI*N$$

where n is the number of modes with that symmetry, h is the total number of symmetry operations, X_I is the number of irreducible representations (the value calculated above), X_R is the number of reducible representations (the entry in the character table), and N is the number of identical symmetry operations. Finally, subtracting the rotational modes and translational modes from the reduced representation gives the number of vibrational modes.

Infrared Spectroscopy

A vibrational mode will be observed in an infrared spectrum if it leads to a change in the molecular dipole moment. Compared to Raman spectroscopy, the infrared photon is completely absorbed and its energy is transferred to the vibration of the molecule, not re-emitted.

Different bonds have different energies associated with them, and require different amounts of energy to stretch or bend. In general, the stronger the bond, the more energy required to deform it. So very weak bonds will only be deformed by low energy radiation, and strong bonds will only be deformed by high energy radiation. This leads to characteristic frequencies where only certain vibrations are absorbed. For example, C-H bonds are typically the only bonds observed in the range from 2960-2850 cm^{-1}. So if absorption is present at that frequency, it can be assumed that it is due to a C-H bond.

Raman Spectroscopy

A vibrational mode will be observed in Raman spectroscopy if it leads to a change in the polarizability of the electron cloud of a molecule. If a photon has energy that is significantly higher than the energy of the vibrational states, it may either be deflected without any change in energy, or it may interact with the molecule and either takes energy from it or give energy to it.

When the photon is absorbed and re-emitted in a different direction, it is called

Rayleigh scattering, and this is strongly dependent on the wavelength of the incoming light. When there is a change in the energy of the photon, it is called Stokes scattering or anti-Stokes scattering, depending on whether energy is absorbed or lost by the molecule. This change in the behavior of the incoming photons can be measured, and will provide information about the concentration and chemical properties of the analyte.

References

- X-ray, science: britannica.com, Retrieved 24, April 2020
- A-Vibrational-Spectroscopy, A-Introduction-to-molecular-symmetry: libretexts.org, Retrieved 17, August 2020,

6

Diverse Aspects of Bioinorganic Chemistry

There are various aspects which are studied within the domain of bioinorganic chemistry. A few of them are biological nitrogen fixation, molybdenum nitrogenase, bioenergetics and glucose storage. The topics elaborated in this chapter will help in gaining a better perspective about these aspects of bioinorganic chemistry.

Biological Nitrogen Fixation

Nitrogen is an essential element contained in many biomolecules, and is therefore necessary for growth and development of the plant. This element is abundantly available in the earth's atmosphere in the form of dinitrogen (N_2) gas, yet most organisms are unable to metabolize N2 because it is relatively inert.

Why is Nitrogen inert? Because it has:

- High Ionization Potential (15.6 eV).
- Negative Electron Affinity (-1.6 eV).
- High Bond Dissociation Energy (225 Kcal/mol).
- Large energy difference of 22.9 eV between HOMO & LUMO.
- Filled MO are lower in energy (≤ -15.6 eV) and its vacant orbitals are high (≥ -7 eV).
- It is Very difficult to add or remove electrons from dinitrogen molecule in it's the ground state.

Need for Nitrogen Fixation

Nitrogen is a critical limiting element which is used for the growth and production of plants. It is a major constituent of chlorophyll, the most imperative pigment needed to carry out photosynthesis. Likewise it is a vital component in amino acids, which are the key building units of proteins. Besides this, it is also present in other vital biomolecules, such as ATP and nucleic acids.

Even though it is one of the most available elements, plants cannot extract nitrogen directly from the atmosphere and can only utilize reduced forms of this element. Therefore, nitrogen fixation, whether natural or synthetic, is necessary for all forms of life. Moreover, it is necessary for agriculture and the production of fertilizer. It is also an essential process in the production of explosives (e.g. gunpowder, dynamite, TNT, etc.).

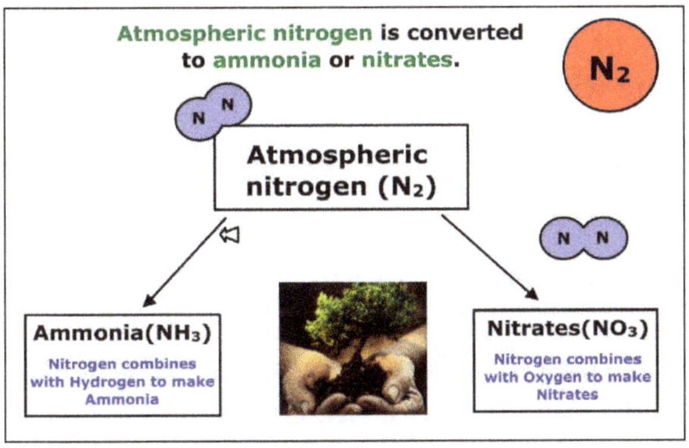

Nitrogen Fixation.

Processes Responsible for Nitrogen Fixation in Biosphere

Typically, all the nitrogen fixation happening in the biosphere can be attributed to three processes:

- Atmospheric fixation.
- Industrial fixation.
- Biological fixation.

Atmospheric Nitrogen Fixation

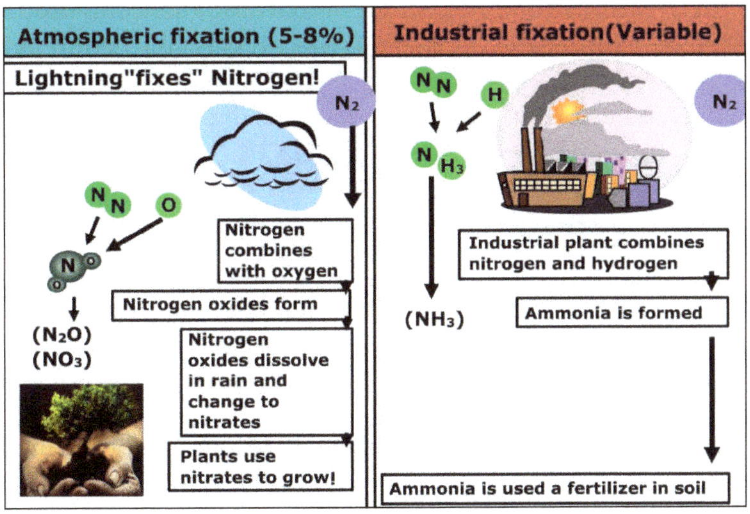

Non-Biological Nitrogen Fixation.

Non-biological nitrogen fixation can happen by electric discharge (lightning) in the troposphere, and by cosmic radiation in the stratosphere.

- Equation involved: $N_2 \rightarrow 2N$; $N + O_2 \rightarrow NO + O \rightarrow NO_x$
- It accounts for only 10 % of N_2 fixation.
- The high energy of lightening breaks N_2 molecules apart & aids the nitrogen atoms to combine with O_2 forming nitrogen oxides.
- They dissolve in rain water forming nitrates and are carried to the ground with the rain and then plants use nitrates to grow.

Industrial Nitrogen Fixation

- The Haber-Bosch process: This process directly synthesizes ammonia from nitrogen and hydrogen. In 1909, the German chemist named Fritz Haber ascertained that atmospheric nitrogen could be combined with hydrogen under extremely high temperature and pressure condition which is catalyzed by an iron catalyst to yield an extremely high proportion of ammonia, which is the starting point for the production of a wide range of nitrogen compounds. This process was made commercially feasible by Carl Bosch and now called as the Haber-Bosch method or the synthetic ammonia process. The Haber-Bosch process is now one of the largest and most-basic processes of the chemical industry throughout the world.
- $N_2 + 3H_2 \rightleftharpoons 2NH_3$ (only 20% conversion).
- NH_3 so produced can be used directly as fertilizer, but most of it is further processed to urea and ammonium nitrate (NH_4NO_3).
- Besides this, combustion of fossil fuels (natural gas, coal, crude oil) and products produced from crude oil (petrol, diesel, gasoline) contributes to nitrogen fixation.

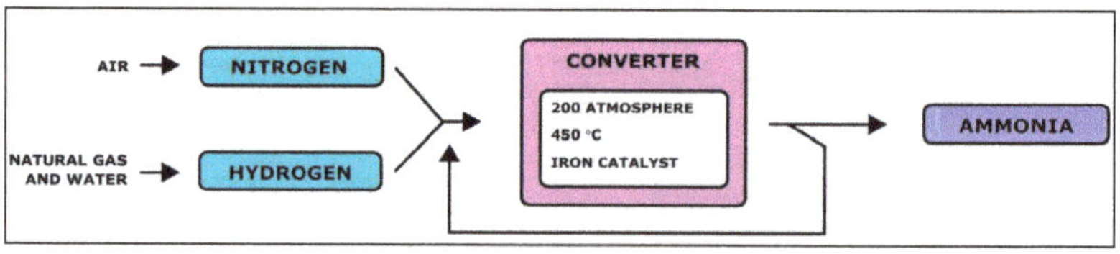

The Haber - Bosch process.

Problems Associated with Industrial Nitrogen Fixation

There are many consequences of using industrial nitrogen fixation which are as follows:

- Industrial production of NH_3 is expensive.

- Transportation and distribution of fertilizer is labor intensive.
- Application of nitrogen on the soil usually causes contamination of adjacent water sources.
- Use of fossil fuels for the energy required to produce this fertilizer results in emission of carbon dioxide contributing to global warming.
- Excessive use of these chemical fertilizers upsets the nitrogen cycle and consequently led to surface water as well as groundwater pollution, thereby harming ecosystem.
- Enlarged heaps of nitrogen fertilizers to fresh water, as well as marine ecosystems, have caused eutrophication, the process whereby these systems cause a proliferation of microorganisms, particularly algae.

Biological nitrogen fixation plays an essential part in providing nitrogen for other forms of life on earth, since it contributes about 60% of the total N_2 fixed in the biogeochemical nitrogen cycle. BNF is therefore called a key for sustenance of agriculture and reduction of soil fertility decline. In 1901, Beijerinck discovered the BNF which is carried out by a specific group of prokaryotes. These organisms utilize an enzyme called nitrogenase to catalyze the conversion of atmospheric nitrogen (N_2) to ammonia (NH_3). Plants can conveniently absorb NH_3 to synthesize the aforementioned nitrogenous biomolecules. This biological process though less efficient, occurs at room temperature and pressure. Therefore, the natural process of BNF deals an economic means for the reduction of environmental problems and improvement of the internal resources.

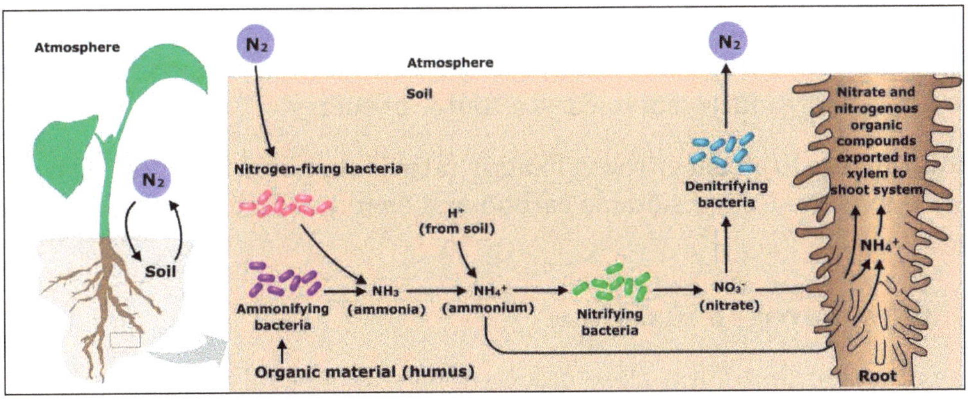

Biological nitrogen fixation.

In this process microorganisms transform atmospheric nitrogen to ammonia, assailable by associated plants. These microorganisms need 16 moles of adenosine triphosphate (ATP) for the reduction of each mole of nitrogen and they obtain this energy by oxidation of organic molecules. Free-living microorganisms which are non-photosynthetic in nature, receive these molecules from other organisms, while photosynthetic microorganisms, such as cyanobacteria, use sugars synthesized in the process of photosynthesis.

$$N_2 + 8H^+ + 8e + 16\ Mg\text{-}ATP \rightarrow 2NH_3 + H_2 + 16\ Mg\text{-}ADP + 16\ P_i$$

Associative and symbiotic nitrogen-fixing microorganisms obtain these compounds from their host plants' rhizospheres.

- Asymbiotic Bacteria or free living bacteria- fixes 30 % of N_2.
- Symbiotic Bacteria- fixes 70 % of N_2 & live in a symbiotic relationship with plants of the legume family (e.g., alfalfa, soybeans).

Biological Nitrogen Fixation (BNF)

Two kinds of nitrogen-fixing bacteria are known: free-living or non-symbiotic bacteria, including the cyanobacteria (or blue-green algae) Anabaena and Nostoc and genera such as Azotobacter,Beijerinckia, and Clostridium; and mutualistic or symbiotic bacteria such as Rhizobium, associated with leguminous plants (e.g., various members of the pea family), and certain Azospirillum species, associated with cereal grasses.

Nitrogen Fixation by Free-Living Bacteria

There are many heterotrophic bacteria which reside in ground soil and are able for fixation of significant levels of nitrogen without the direct interaction with other beings. Examples for this type of nitrogen-fixing bacteria include species of Azotobacter, Bacillus, Clostridium, and Klebsiella. These organisms search their own source of energy either by oxidation of organic molecules released by other organisms or from decomposition. Some free-living organisms have chemolithotrophic capabilities which help them to utilize inorganic compounds as a source of energy.

Their contribution to global nitrogen fixation rates is supposed to be less than generally observed due to the lack of suitable carbon and energy sources for these microorganisms.

Associative Nitrogen Fixation

Species of Azospirillum form close associations with several members of the Poaceae (grasses), that includes agriculturally economic and vital cereal crops, such as rice, barley, oats, wheat and corn. These bacteria fix appreciable amounts of nitrogen within the rhizosphere of the host plants. The degree of N fixation is limited by various elements, including soil temperature (Azospirillum species thrive in more temperate and/or tropical environments), the power of the host plant to provide a rhizosphere environment low in oxygen pressure, the availability of host photosynthates for the bacteria, the competitiveness of the bacteria, and the efficiency of nitrogen.

Symbiotic Nitrogen Fixation

Many microorganisms fix nitrogen symbiotically by partnering with a host plant. Sugars are produced by plants via photosynthesis that are utilized by the nitrogen-fixing microorganism for the energy it required for nitrogen fixing. In exchange for these carbon sources, the microbe provides fixed nitrogen to the host plant for its growth. Water fern Azolla's which symbiosis with a cyanobacterium Anabaena azollae is a type of example for this type of nitrogen fixation. Even though the symbiotic partners described above play a vital role in the worldwide ecology of nitrogen fixation, till date relationships between legumes and Rhizobium and Bradyrhizobium bacteria are considered to be the most important nitrogen-fixing symbiotic associations.

Legume Nodule Formation

The symbiotic nitrogen-fixing bacteria invade into the host plant's root system, where they multiply and stimulate the formation of root nodules, enlargements of plant cells and bacteria in intimate association. The bacteria then start to fix the nitrogen required by the plant present in the nodules. Access to the fixed nitrogen helps the plant to synthesize leaves fortified with nitrogen which can be recycled throughout the plant. This results in increased photosynthetic capacity, which in turn yields nitrogen-rich seed. It is quite amazing to note the outcomes of legumes which have not being nodulated, especially when nitrogen poor soil is used for the plant growth. The plants produced are usually low in nitrogen content, and yield very little seed. Table consists of the crops with their respective nodulating bacteria.

Table: Major cross inoculation groups of legumes.

Crop	Nodulating Bacteria
Alfalfa	Sinorhizobium melilotii
Beans	Rhizobium legumninosarum biovar phaseoli and Rhizobium tropici
Clover	Rhizobium leguminosarum biovar trifolii
Lotus	Mesorhizobium loti
Peas	Rhizobium leguminosarum biovar viceae
Soybean	Bradyrhizobium japonicum, Bradyrhizobium elkanii, Rhizobium fredii
Sesbania	Azorhizobium caulinodans

Benefits of Nitrogen Fixation

- Atmospheric nitrogen is converted to ammonium ion and other nitrogen compounds that can be used by plants.
- It results in increase of soil fertility by nitrates and nitrites.

- Nitrogen fixing organisms increase the humus content of the soil and water holding capacity of soil.
- Beneficial organisms like rhizobium, cyanobacteria, blue-green algae and fungi increase their number in soil.
- Plants avail nitrogen by soil and animals can obtain required nitrogen from plants. That's how; nitrogen from abiotic pool is incorporated in biotic pool.

Factors Limiting Biological Nitrogen Fixation

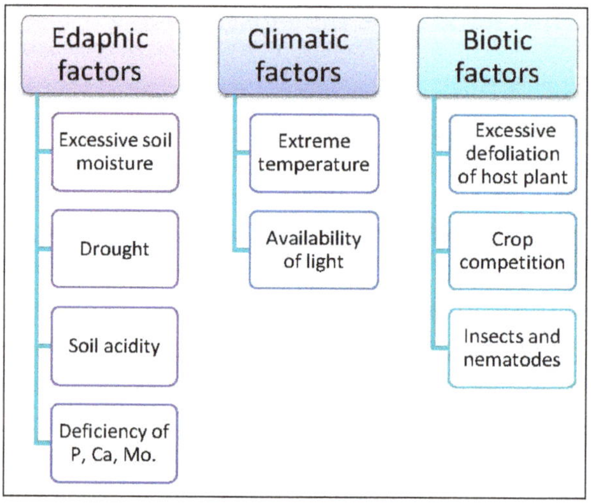

Nitrogen Cycle Maintains a Pool of Biologically Available Nitrogen

Biologically available nitrogen is restored and reused during the metabolic processes of different species in the process called nitrogen cycle. The first step in the cycle is fixation (reduction) of atmospheric nitrogen by nitrogen-fixing bacteria to yield ammonia (NH_3 or NH_4^+).

This ammonia can be used by most living organisms, but soil bacteria derive their energy by oxidizing ammonia to nitrite (NO_2^-) and ultimately nitrate (NO_3^-). These bacteria are so abundant and active that nearly all ammonia reaching the soil is oxidized to nitrate. The chemical process is known as nitrification.

The nitrates and nitrites so formed is taken up and readily reduced to ammonia by plants and bacteria through the action of nitrate and nitrite reductases. This ammonia is incorporated into amino acids by plants and animals then use plants as a source of amino acids, to build their proteins. When organisms die, microbial degradation of their proteins returns ammonia to the soil, where nitrifying bacteria convert it back to nitrite and nitrate. Thus, a balance is maintained between fixed nitrogen and atmospheric nitrogen by bacteria that convert nitrate to N_2 under anaerobic conditions, a process called denitrification.

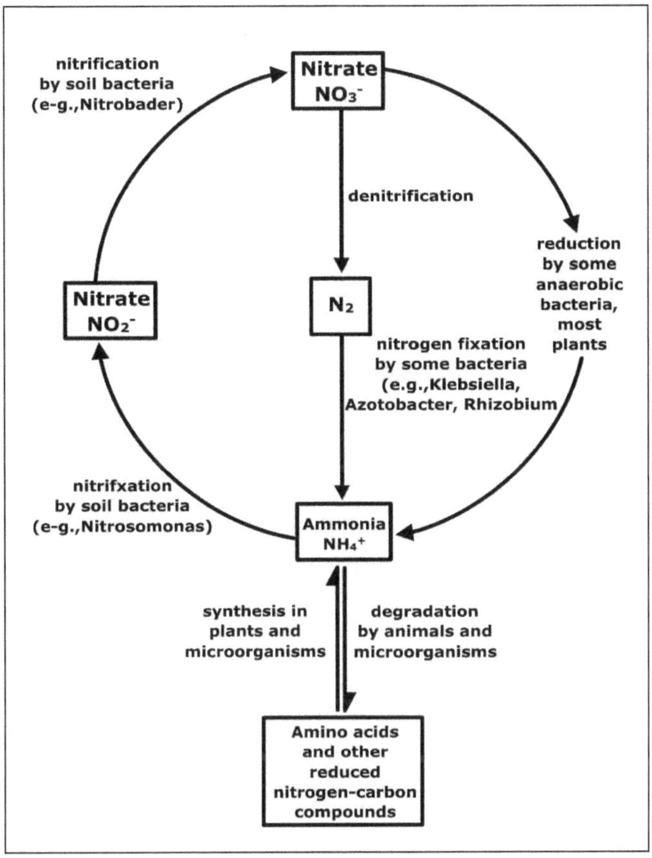

The Nitrogen Cycle.

Molybdenum Nitrogenase

The enzyme nitrogenase is found in certain bacteria and blue-green algae, which can reduce N_2 to NH_3 (nitrogen fixation). Some of these bacteria are free-living while others are symbiotic (in the anaerobic environment of roots of legume plants). This bacterial reaction is the key step in the nitrogen cycle, which maintains a balance between two reservoirs of the nitrogen compounds. Nitrogen-fixing bacteria catalyze dinitrogen reduction in forming two ammonia molecules, resulting the major contribution of fixed nitrogen to the biogeochemical nitrogen cycle.

Although, the reduction of dinitrogen still remains one of the today's biggest challenges in chemistry. As a result, substantial efforts have been applied to understanding the mechanism of bacterial nitrogenases. The transient interaction of two component proteins are involved in the reduction of N_2 by this enzyme, designated the Fe protein and the MoFe protein, and minimally requires sixteen MgATP, eight protons and electrons.

Four types of nitrogenases are known till date, each having a different combination of metals at the active site But, the most abundant nitrogenase type is the Monitrogenase, which has an active site called FeMo-cofactor This cofactor contains Mo along with Fe, S, (R)-homocitrate, and a light atom of unknown identity called X.

The ideal reaction catalyzed by the Mo-dependent nitrogenase can be represented by the chemical equation:

$$N_2 + 8e^- + 8H^+ + 16\ ATP + 16\ H_2O \rightleftharpoons 2NH_3 + H_2 + 16\ ADP + 16\ P_i$$

The Mo-nitrogenase is composed of two-component metalloproteins commonly called as the iron protein and the molybdenum-iron protein.

Components of Mo-Nitrogenase

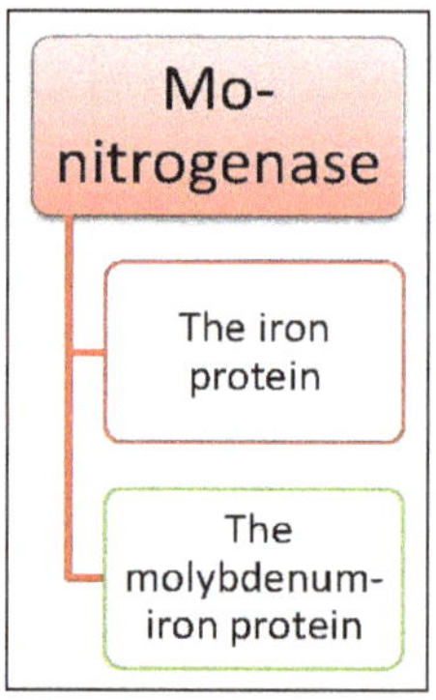

Their structures either individually or in the complex form are determined by X-ray crystallography that has provided detailed models of the three metal clusters present in nitrogenase. The nitrogen fixation by nitrogenase contains a complex interaction between the two protein components, MgATP electrons and protons. Understanding the molecular details of this multi-step and multi-component system is difficult.

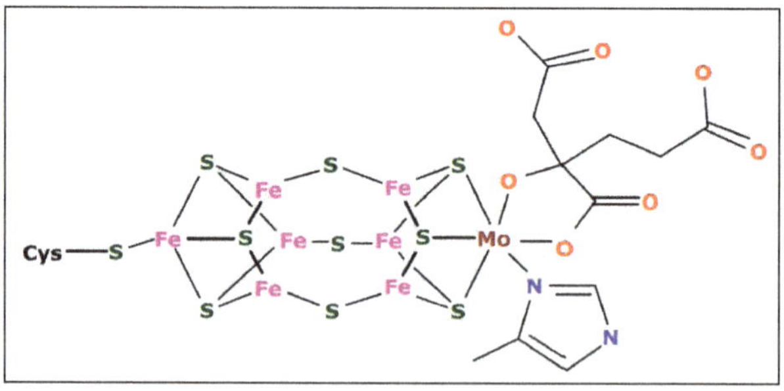

Structure of M-cluster of Mo-nitrogenase.

Iron Protein (Fe Protein)

Fe Protein Cycle

The Fe protein is a homodimer (Molecular weight 64,000) consisting one nucleotide

(MgATP/MgADP) binding site within each subunit and a single [4Fe-4S] cluster bridging the two subunits. The role of Fe protein in nitrogenase is to transport electrons (one at a time) to the MoFe protein in a process coupled with the hydrolysis of two MgATP molecules. In nitrogenase mechanism, Fe protein participated can be considered to operate in a three state cycle (known as the Fe protein cycle). The reduced Fe protein in the cycle, with its [4Fe-4S] cluster in the 1^+ oxidation state, has two MgATP binding sites. It is this state that transiently associates with the MoFe protein.

While associating, the two MgATP molecules bound to Fe protein are hydrolyzed to two MgADP molecules, and a single electron is transferred from the Fe protein [4Fe-4S] cluster into the MoFe protein.

The oxidized Fe protein ($[4Fe\text{-}4S]^{2+}$) with two bound MgADP molecules, further gets dissociated from the MoFe protein.

The free Fe protein is regenerated then in two steps. Again reduction of $[4Fe\text{-}4S]^{2+}$ cluster takes place resulting 1^+ oxidation state and MgADP molecules are replaced by MgATP molecules.

Physiologically, reduction of Fe protein depends upon the organism, with reduced ferredoxin or flavodoxin being the most common electron donor. Lots of questions still remain unanswered regarding the mechanistic details of the Fe protein cycle. For instance, the nature of the interaction between the nucleotide binding sites in the Fe protein and MoFe protein interface remain obscure. The reason behind undetectable display rates for MgATP hydrolysis in Fe protein prior to association with the MoFe protein. How the hydrolysis of MgATP gets activated, once it bound to the MoFe protein. Likewise, very little is known about how MgATP binding and hydrolysis within the Fe protein is communicated into the MoFe protein. This type of communication is suggested by the observation that the Fe protein is the only known reductant for the MoFe protein that will support substrate reduction. The capacity of small-molecule electron donors to reduce the MoFe protein without an related ability to support substrate reduction demonstrates that the Fe protein induce changes within the MoFe protein and should be coupled to MgATP hydrolysis.

MoFe-Protein

The MoFe protein is a heterotetramer (Mr≈ 250,000) conatining two pairs of metalloclusters, named P-cluster ([8Fe-7S]) and FeMo-cofactor ([7Fe-Mo-9S-homocitrate-X]).

FeMo-Cofactor

FeMo-cofactor consists of a transition metal-sulfur framework and one molecule of (R)- homocitrate. It is the site of substrate binding and reduction.

A [4Fe-3S] subcluster is connected to a [3Fe-Mo-3S] subcluster by an atom X at one corner and three bridging inorganic sulfides. (R)- Homocitrate is coordinated to the Mo atom through its 2-hydroxy and 2-carboxyl groups. At the centre of FeMo-cofactor, the presence of the light atom (X) was detected in a high-resolution (1.16 Å) structure of the MoFe protein. The electron density for X is mostly consistent with N, C, or O, but efforts to identify this atom have yet to be successful.

FeMo-cofactor is anchored to the MoFe protein by α-275Cys to an iron atom at one end and α442His to the Mo atom at the other end.

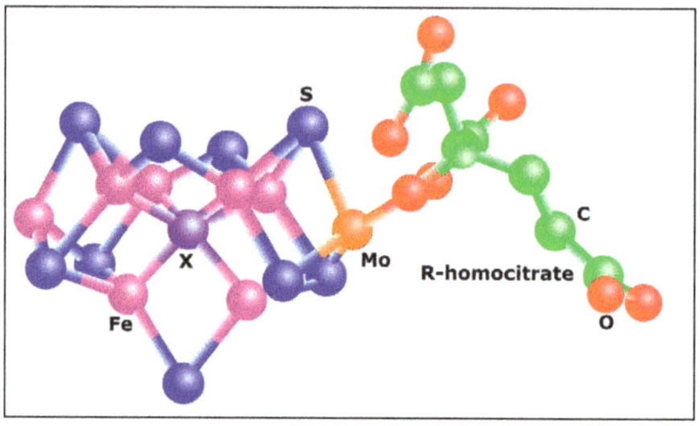

Structure of MoFe-cofactor.

There is considerable evidence that the FeMo-cofactor is the site of substrate binding and reduction:

- Certain mutant strains that are deficient in FeMo-cofactor biosynthesis are inactive for reduction of all substrates, but gets activated on the addition of an isolated FeMo-cofactor.
- MoFe protein that has a structurally altered FeMo-cofactor, wherein citrate substitutes for homocitrate, display altered catalytic properties.
- Amino acid substitutions in the FeMo-cofactor binding environment bring out changes in the catalytic and spectroscopic properties of the altered MoFe protein.
- In CO-inhibited MoFe protein, ENDOR spectroscopy demonstrates that the CO binds to FeMo-cofactor, not to the P cluster.

Role of Molybdenum

- It affects the reactivity of the cofactor involved in nitrogen fixation.
- The Mo atoms slow down the protonation of the cluster at the active site, that suppresses the production of dihydrogen.
- It also maximizes the possibility of dinitrogen binding to that active site.
- Mo-Fe-S cluster protonate slowly than Fe-S cluster and it has also been shown that Mo containing Fe-S clusters have greater substrate affinity as compared to Fe-S clusters.

- It modulate the reduction potential necessary for N_2 binding and reduction.
- It participates in electron transfer between the P-cluster pair and FeMo cofactor.
- It plays a structural role by providing a polydentate site that can simultaneously interact with the remainder of the cofactor, the homocitrate, and the protein, thereby serving as a structural anchor to maintain all these groups in their proper orientations.
- Thus, the role of Mo as cofactor is justified and it has been used for in-vitro nitrogen fixation as well.

P-Cluster

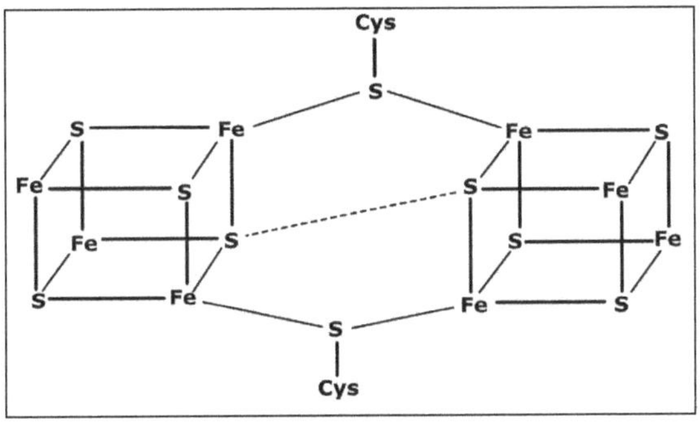

P-cluster structures.

These are [8Fe-7S] clusters; they are responsible for electron transfer between the Fe protein and the substrate reduction site FeMo-cofactor due to the following:

- According to the X-ray crystal structures of two different stable Fe protein and MoFe-protein complexes, P-cluster is placed in between the Fe protein cluster and FeMo-cofactor.
- Amino acid substitutions placed between the P-cluster and the MoFe-cofactor within the MoFe protein disturb the intramolecular transfer of electrons from the P-cluster to FeMo-cofactor.
- Stopped-flow and EPR spectroscopic results obtained under turnover conditions show that during the reduction of bound dinitrogen, the P-cluster is oxidized.
- Fe protein appreciates the MgATP-dependent transfer of an electron to the oxidized P-cluster of the MoFe protein.
- As indicated by freeze-quench EPR spectroscopy, the P-clusters within a MoFe protein having the β-188Ser residue substituted by cysteine change oxidation state during turnover.

The P-cluster are made from two linked [4Fe-4S] subclusters having common bridging μ6-sulfide at one corner. Cysteinate-S ligands is the only known naturally occurring [Fe-S] cluster containing serinate-O (β-188Ser) and amideN (α-88Cys) ligands. X-ray crystallographic studies of two different oxidation states of the MoFe protein tells that P-clusters undergo structural rearrangement upon oxidation (pOX) from the resting state (dithionite reduced; pN), one consequence of which is the displacement of both the serinate-O and amideN ligands from pN.

There is very less information about the oxidation states that the P-cluster accesses during turnover. According to this information a practical working model of P-cluster function is made. According to which, firstly one or two electron is transferred from the P-cluster to FeMo-cofactor upon association of the Fe protein with the MoFe protein, that is followed by re-reduction of the oxidized P-cluster by electron transfer from the Fe protein.

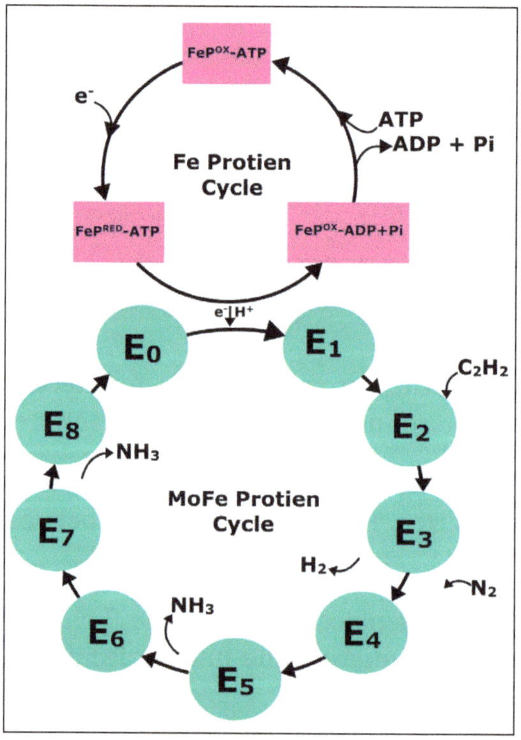

Fe and MoFe-protein catalytic cycles.

During catalysis, the FeMo-cofactor should accept eight electrons, whereas only one electron is delivered during each event of Fe protein binding.

The subscript denotes the number of electrons (and protons) added to the MoFe protein (E).

Thus, MoFe protein proceeds through states from E_0 to E_8 during N_2 fixation before it returns back to the resting state (E_0).

The 1-electron Fe protein cycle and 8-electron MoFe-protein cycle can be conceived of as interlocking, with the Fe protein cycle driving the MoFe protein to successively reduced states.

This model for the nitrogenase mechanism depicts several important observations regarding the mechanics of catalysis.

For instance, it is known that three or four electrons must accumulate within the MoFe protein before N_2 binds (E_3 or E_4 states).

Moreover, a stoichiometric quantity of H_2 is evolved, when N_2 binds to the MoFe protein. Besides, reducing N_2 and protons, nitrogenase has the ability to reduce a wide variety of small unsaturated molecules, such as azides, nitrous oxide, nitriles, isonitriles, and alkynes.

The reduction of the acetylene (C_2H_2) to ethylene (C_2H_4) is the most commonly used method for monitoring nitrogenase activity.

Even though both acetylene and N_2 have potential to bind to the same site on FeMo-cofactor, it is worth mentioning that acetylene binds to a less-reduced E state (E_2) than does N_2 (E_3, E_4). Therefore, when these two compounds are present, acetylene appears to be a non-competitive inhibitor of N_2 reduction.

Substrate Entry and Product Release

Although the FeMo-cofactor is buried quite below the surface of protein and no permanent channels are there between the MoFe-cofactor and the protein, but there are two clefts which can be used for the substrate entering and release of product and/or transfer of H_3O^+ to the active site.

The first cleft is made up of five stretches of polypeptide chain (a255-a275, a279- a301, a338-a359, a363-a378, and a422-a436). The second cleft is also made up of five stretches of polypeptide chain (a172- a199, a341-a352, a365-a400, a256-a276, and a34-a43).

Both of these clefts have funnel like shape and are near to one of the putative Fe-protein subunit binding sites. However, these two clefts are not wide enough to allow free diffusion of either substrate or product or H_3O^+. Therefore, it seems reasonable to assume that some conformational changes must be there which provides diffusion of ligands into and away from the FeMocofactor, similar to the case of binding and release of oxygen to the buried hemes in globins.

The details of the interaction between various substrates and the FeMo-cofactor could provide understanding in the catalytic function of nitrogenase. Therefore, it becomes necessary to understand different modes of interactions between them. There are many possibilities by which substrate can bind to one or more of the Fe, Mo, and/or S site. Of all, the six trigonally coordinated iron sites of the FeMo-cofactor are particularly fascinating for the substrate binding.

Based on the FeMo-cofactor structure, three different types of binding modes of substrates to these coordinatively unsaturated irons may be envisioned:

- End-on fashion: Substrates could bind in an end-on fashion to one of the six central iron sites which are bridged by non-protein ligands. It is also conceivable that some substrates could displace the bridging ligand Y, if this ligand is not sulfur, and interact simultaneously with the two adjacent iron sites.
- The central cavity: Some small substrates, such as N_2 and/or H^+, could occupy the central cavity in the FeMo-cofactor, thereby replacing the weak iron-iron bonds with Fesubstrate bonds.
- The exterior surface: Substrates could bind to the exterior surface of the FeMo-cofactor. Three sets of cyclic eight-membered rings occur on the exterior surface of the FeMocofactor, consisting of alternating S(Y) and Fe sites.

The existence of multiple potential substrate-binding sites in the FeMo-cofactor could be linked with the noncompetitive kinetics observed between N_2 and other substrates. The complete structural model for nitrogen fixation is shown in figure.

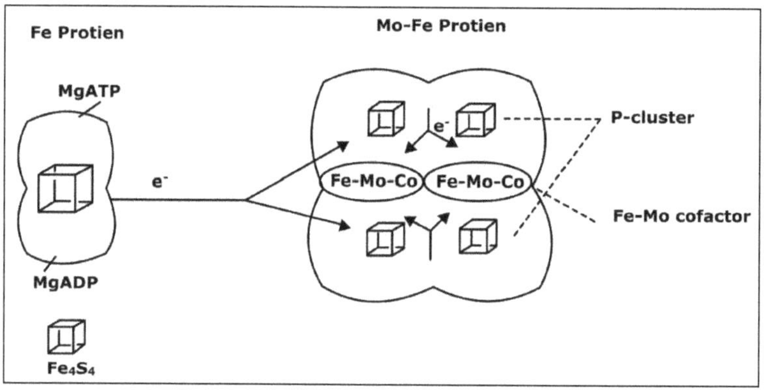

Complete Structural Model for Nitrogen fixation.

Bioenergetics

Bioenergetics means study of the transformation of energy in living organisms. The goal of bioenergetics is to describe how living organisms acquire and transform energy in order to perform biological work. The study of metabolic pathways is thus essential to bioenergetics. In a living organism, chemical bonds are broken and made as part of the exchange and transformation of energy. Energy is available for work (such as mechanical work) or for other processes (such as chemical synthesis and anabolic processes in growth), when weak bonds are broken and stronger bonds are made. The production of stronger bonds allows release of usable energy. Adenosine triphosphate (ATP) is the main "energy currency" for organisms; the goal of metabolic and catabolic processes are to synthesize ATP from available starting materials (from the environment), and to break-down ATP (into adenosine diphosphate (ADP) and inorganic phosphate) by utilizing it in biological processes. In a cell, the ratio of ATP to ADP concentrations is known as the "energy charge" of the cell. A cell can use this energy charge to relay information about cellular needs; if there is more ATP than ADP available, the cell can use ATP to do work, but if there is more ADP than ATP available, the cell must synthesize ATP via oxidative phosphorylation. Living organisms produce ATP from energy sources via oxidative phosphorylation. The terminal phosphate bonds of ATP are relatively weak compared with the stronger bonds formed when ATP is hydrolyzed (broken down by water) to adenosine diphosphate and inorganic phosphate. Here it is the thermodynamically favorable free energy of hydrolysis that results in energy release; the phosphoanhydride bond between the terminal phosphate group and the rest of the ATP molecule does not itself contain this energy.

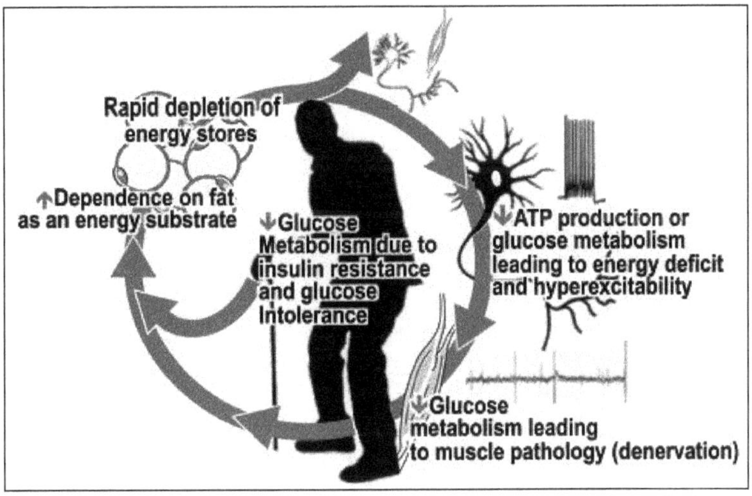

Bioenergetic deficit leading to neuronal death and muscle pathology.

Types of Bioenergetics Reactions

Exergonic Reaction

Exergonic implies the release of energy from a spontaneous chemical reaction without any concomitant utilization of energy. The reactions are significant in terms of biology as these reactions have an ability to perform work and include most of the catabolic reactions in cellular respiration. Most of these reactions involve the breaking of bonds during the formation of reaction intermediates as is evidently observed during respiratory pathways. The bonds that are created during the formation of metabolites are stronger than the cleaved bonds of the substrate. The release of free energy, G, in an exergonic reaction (at constant pressure and temperature) is denoted as:

$$\Delta G = G_{products} - G_{reactants} < 0$$

Endergonic Reactions

Endergonic in turn is the opposite of exergonic in being non-spontaneous and requires an input of free energy. Most of the anabolic reactions like photosynthesis and DNA and protein synthesis are endergonic in nature. The release of free energy, G, in an exergonic reaction (at constant pressure and temperature) is denoted as:

$$\Delta G = G_{products} - G_{reactants} > 0$$

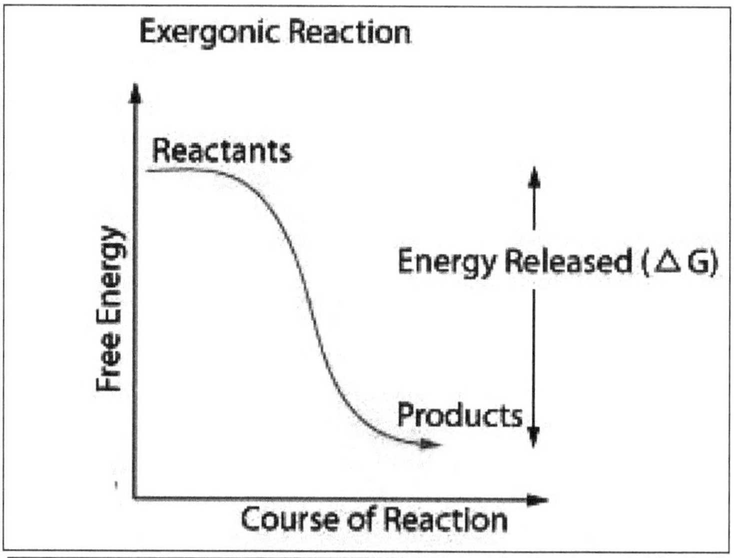

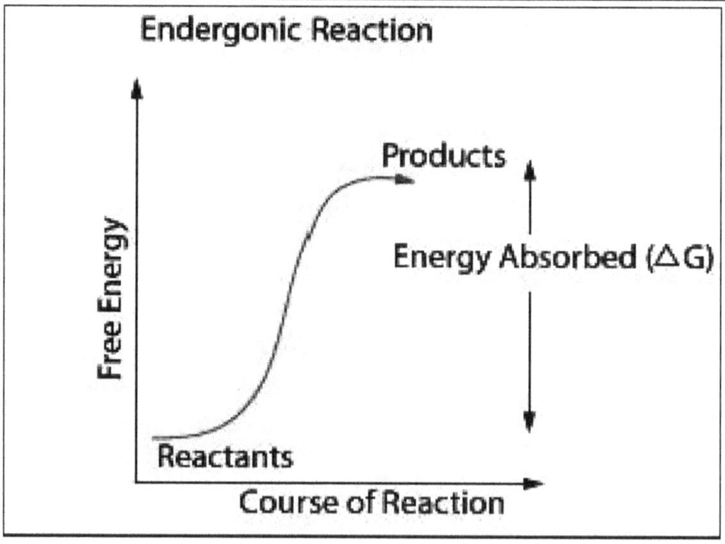

Activation Energy

Activation energy is the energy which must be available to a chemical system with potential reactants to result in a chemical reaction. Activation energy may also be defined as the minimum energy required starting a chemical reaction.

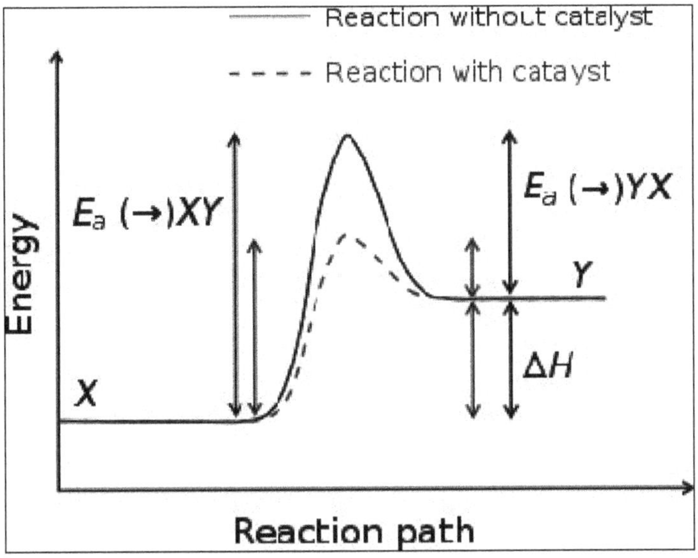

Examples of Major Bioenergetics Processes

- Glycolysis is the process of breaking down glucose into pyruvate, producing net eight molecules of ATP (per 1 molecule of glucose) in the process. Pyruvate is one product of glycolysis, and can be shuttled into other metabolic pathways (gluconeogenesis, etc.) as needed by the cell. Additionally, glycolysis produces equivalents in the form of NADH (nicotinamide adenine dinucleotide), which will ultimately be used to donate electrons to the electron transport chain.
- Gluconeogenesis is the opposite of glycolysis; when the cell's energy charge is low (the concentration of ADP is higher than that of ATP), the cell must synthesize glucose from carbon- containing biomolecules such as proteins, amino acids, fats, pyruvate, etc. For example, proteins can be broken down into amino acids, and these simpler carbon skeletons are used to build/ synthesize glucose.
- The citric acid cycle is a process of cellular respiration in which acetyl coenzyme A, synthesized from pyruvate dehydrogenase, is first reacted with oxaloacetate to yield citrate. The remaining eight reactions produce other carbon- containing metabolites. These metabolites are successively oxidized, and the free energy of oxidation is conserved in the form of the reduced coenzymes FADH2 and NADH. These reduced electron carriers can then be re- oxidized when they transfer electrons to the electron transport chain.
- Ketosis is a metabolic process whereby ketone bodies are used by the cell for energy (instead of using glucose). Cells often turn to ketosis as a source of energy when glucose levels are low; e.g. during starvation.

- Oxidative phosphorylation and the electron transport chain is the process where reducing equivalents such as NADPH, $FADH_2$ and NADH can be used to donate electrons to a series of redox reactions that take place in electron transport chain complexes. These redox reactions take place in enzyme complexes situated within the mitochondrial membrane. These redox reactions transfer electrons "down" the electron transport chain, which is coupled to the proton motive force. This difference in proton concentration between the mitochondrial matrix and inner membrane space is used to drive ATP synthesis via ATP synthase.
- Photosynthesis, another major bioenergetic process, is the metabolic pathway used by plants in which solar energy is used to synthesize glucose from carbon dioxide and water. This reaction takes place in the chloroplast. After glucose is synthesized, the plant cell can undergo photophosphorylation to produce ATP.

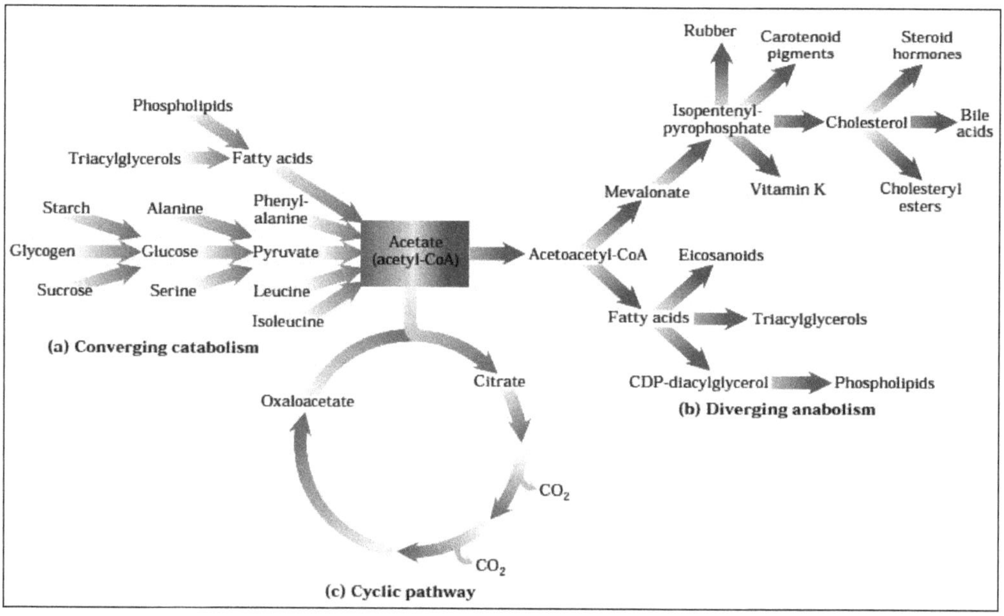

Bioenergetics Relationship Between Free Energy, Enthalpy and Entropy

Every living cell and organism must perform work to stay alive, to grow and to reproduce. The ability to harvest energy from nutrients or photons of light and to channel it into biological work is the miracle of life.

- 1st Law of Thermodynamics: The energy of the universe remains constant.
- 2nd Law of Thermodynamics: All spontaneous processes increase the entropy of the universe.

The important state functions for the study of biological systems are:

- The Gibbs free energy (G) which is equal to the total amount of energy capable of doing work during a process at constant temperature and pressure.
    - If ΔG is negative, then the process is spontaneous and termed exergonic.
    - If ΔG is positive, then the process is nonspontaneous and termed endergonic.
    - If ΔG is equal to zero, then the process has reached equilibrium.
- The Enthalpy (H) which is the heat content of the system. Enthalpy is the amount of heat energy transferred (heat absorbed or emitted) in a chemical process under constant pressure.
    - When ΔH is negative the process produces heat and is termed exothermic.
    - When ΔH is positive the process absorbs heat and is termed endothermic.
- The Entropy (S) is a quantitative expression of the degree of randomness or disorder of the system. Entropy measures the amount of heat dispersed or transferred during a chemical process.
    - When ΔS is positive then the disorder of the system has increased.
    - When ΔS is negative then the disorder of the system has decreased.

The conditions of biological systems are constant temperature and pressure. Under such conditions the relationships between the change in free energy, enthalpy and entropy can be described by the expression where T is the temperature of the system in Kelvin.

$$\Delta G = \Delta H - T\Delta S$$

[ΔG = Gibbs Free Energy; ΔH = Change in Enthalpy; T = Temperature in K; ΔS = Change in Entropy].

Three Thermodynamic Quantities			
Quantity	**Symbol**	**Measures**	**Units**
Enthalpy	*H*	Heat	Energy
Entropy	*S*	Disorder	Energy/K
Free energy	*G*	Reactivity	Energy

Energy Rich Compounds

High energy phosphates act as energy currency of cell. Three major sources of high energy phosphates taking part in energy conservation or energy capture.

Oxidative Phosphorylation (or OXPHOS)

In metabolic pathway, cells use enzymes to oxidize nutrients, thereby releasing energy which is used to produce adenosine triphosphate (ATP). In most eukaryotes, this takes place inside mitochondria. Almost all aerobic organisms carry out oxidative phosphorylation. This pathway is probably so pervasive because it is a highly efficient way of releasing energy, compared to alternative fermentation processes such as anaerobic glycolysis. The process that accounts for the high ATP yield is known as oxidative phosphorylation. In glycolysis and the citric-acid cycle generate other products besides ATP and GTP, namely NADH and $FADH_2$. These products are molecules that are oxidized (i.e., give up electrons) spontaneously. The body uses these reducing agents (NADH and FADH2) in an oxidation-reduction reaction:

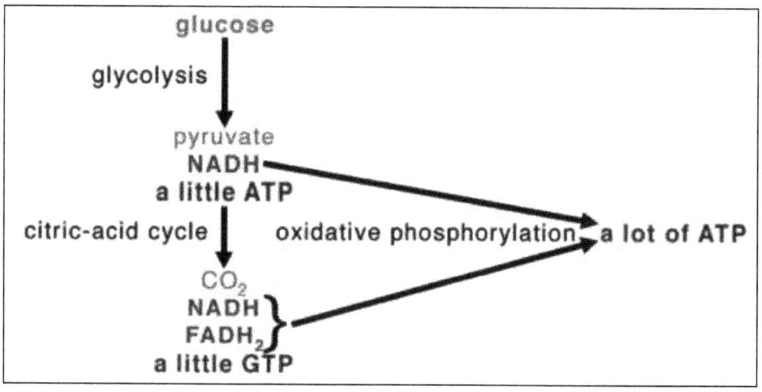

Glycolysis

Cells use the glycolysis pathway to extract energy from sugars, mainly glucose, and store it in molecules of adenosine triphosphate (ATP) and nicotinamide adenine dinucleotide (NADH). The end product of glycolysis is pyruvate, which can be used in other metabolic pathways to yield additional energy. During glycolysis ATP molecules are used and formed in the following reactions (aerobic phase):

Reactions Catalyzed	ATP used	ATP formed
Stage I: 1. Glucokinase (for phosphorylation)	1	
2. Phosphofructokinase I (for phosphorylation)	1	
Stage II: 3. Glyceraldehyde 3-phosphate dehydrogenase (oxidation of 2 NADH in respiratory chain) nresrespiratory chain) respiratory chain)		6

4. Phosphoglycerate kinase (substrate level phosphorylation)		2
Stage IV: 5. Pyruvate kinase (substrate level phosphorylation)		2
Total	2	10
Net gain		08

In the anaerobic phase oxidation of one glucose molecule produces 4 - 2 = 2 ATP.

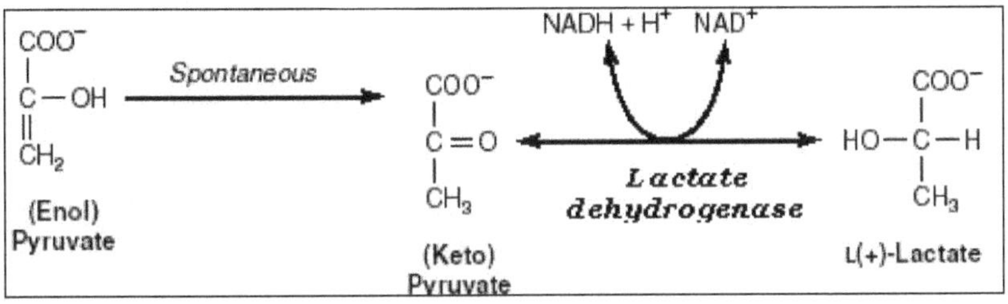

TCA Cycle

The citric acid cycle (CAC) – also known as the tricarboxylic acid (TCA) cycle or the Krebs cycle is a series of chemical reactions used by all aerobic organisms to release stored energy through the oxidation of acetyl- CoA derived from carbohydrates, fats, and proteins into carbon dioxide and chemical energy in the form of adenosine triphosphate (ATP). If one molecule of the substrate is oxidized through NADH in the electron transport chain three molecules of ATP will be formed and through FADH2, two ATP molecules will be generated. As one molecule of glucose gives rise to two molecules of pyruvate by glycolysis, intermediates of citric acid cycle also result as two molecules.

Reactions	No.of ATP formed
1. 2 isocitrate → 2 α-ketoglutarate ($2NADH + 2H^+$) (2 × 3)	6
2. 2 α-ketoglutarate → 2 succinyl CoA ($2NADH + 2H^+$) (2 × 3)	6
3. 2 succinyl CoA → 2 succinate (2 GTP = 2ATP)	2
4. 2 succinate → 2 Fumarate (2 $FADH_2$) (2 × 2)	4
5. 2 malate → 2 oxaloacetate ($2NADH + 2H^+$) (2 × 3)	6
Total No.of ATP formed	24

Energy Shuttles

- NADH: An energy shuttle which delivers high energy electrons to the electron transport chain where they will eventually power the production of 2 to 3 ATP molecules. When this electron shuttle is not carrying high energy electrons, meaning it has been oxidized (lost its electrons), it is left with a positive charge and is called NAD^+.

- $FADH_2$: Another energy shuttle that carries high energy electrons to the electron transport chain, where they will ultimately drive production of 1 to 2 ATP molecules. The oxidized form of $FADH_2$ is FAD and happens just like in NADH.

High Energy Molecules

- ATP: The basic energy currency of the cell. It's a form of energy that cells can use right away.
- GTP: Similar to ATP, GTP can be easily converted to ATP in the cell.

Energy Released by Hydrolysis of Some Phosphate Compounds

Type	Example	Energy Released (kcal/mol)
acyl phosphate 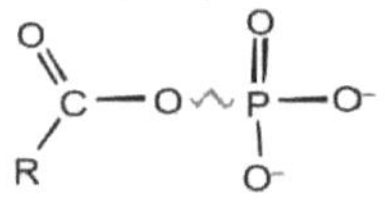	1,3-bisphosphoglycerate (BPG)	−11.8
	acetyl phosphate	−11.3
guanidine phosphates 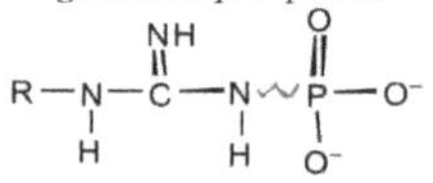	creatine phosphate	−10.3
	arginine phosphate	−9.1
pyrophosphates 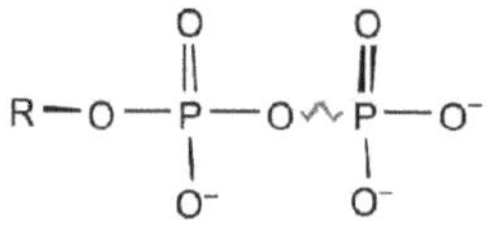	$PP_{i*} \rightarrow 2P_i$	−7.8
	$ATP \rightarrow AMP + PP_i$	−7.7
	$ATP \rightarrow ADP + P_i$	−7.5
	$ADP \rightarrow AMP + P_i$	−7.5
sugar phosphates 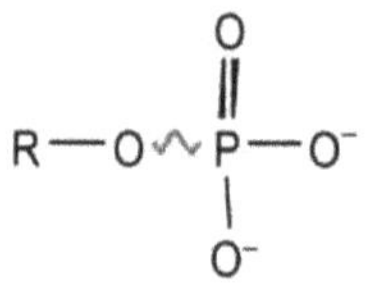	glucose 1-phosphate	−5.0
	fructose 6-phosphate	−3.8
	$AMP \rightarrow adenosine + P_i$	−3.4
	Glucose-6-phosphate	−3.3
	Glycerol-3-phosphate	−2.2
*PP_i is the pyrophosphate ion.		

Adenosine Triphosphate (ATP)

- Adenosine-5'-triphosphate (ATP) is a multifunctional nucleotide used in cells as a coenzyme.
- It is often called the "molecular unit of currency" of intracellular energy transfer. ATP transports chemical energy within cells for metabolism.
- It is produced by photophosphorylation and cellular respiration and used by enzymes and structural proteins in many cellular processes, including biosynthetic reactions, motility, and cell division.

- One molecule of ATP contains three phosphate groups and it is produced by *ATP synthase* from inorganic phosphate and adenosine diphosphate (ADP) or adenosine monophosphate (AMP).
- The structure of this molecule consists of a purine base (adenine) attached to the 1' carbon atom of a pentose sugar (ribose). Three phosphate groups are attached at the 5' carbon atom of the pentose sugar. It is the addition and removal of these phosphate groups that inter-convert ATP, ADP and AMP. When ATP is used in DNA synthesis, the ribose sugar is first converted to deoxyribose by *ribonucleotide reductase*.

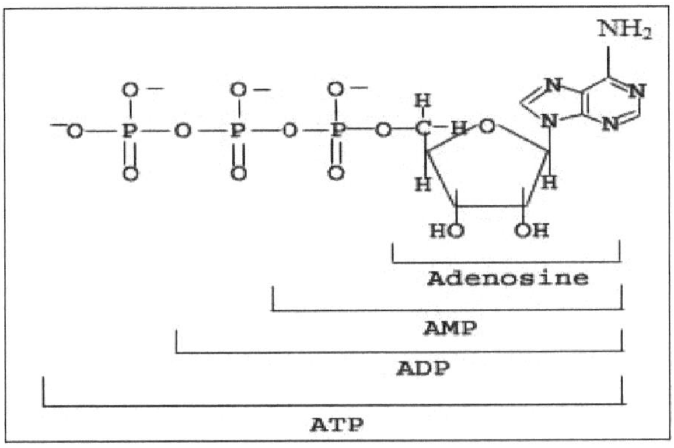

- The three main functions of ATP in cellular function are:
    - Transporting organic substances—such as sodium, calcium, potassium—through the cell membrane.
    - Synthesizing chemical compounds, such as protein and cholesterol.
    - Supplying energy for mechanical work, such as muscle contraction.
- The standard amount of energy released from hydrolysis of ATP can be calculated from the changes in energy under non-natural (standard) conditions, then correcting to biological concentrations. The energy released by cleaving either a phosphate (P_i) or pyrophosphate (PP_i) unit from ATP at standard state of 1 M are:

  $ATP + H_2O \rightarrow ADP + P_i$ $\Delta G° = -30.5$ kJ/mol (−7.3 kcal/mol)

  $ATP + H_2O \rightarrow AMP + PP_i$ $\Delta G° = -45.6$ kJ/mol (−10.9 kcal/mol)

These values can be used to calculate the change in energy under physiological conditions and the cellular ATP/ADP ratio (also known as the Energy Charge). This reaction is dependent on a number of factors, including overall ionic strength and the presence of alkaline earth metal ions such as Mg^{2+} and Ca^{2+}. Under typical cellular conditions, ΔG is approximately −57 kJ/mol (−14 kcal/mol).

Cyclic Adenosine Monophosphate

- It is a second messenger important in many biological processes. cAMP is derived from adenosine triphosphate (ATP) and used for intracellular signal transduction in many different organisms, conveying the cAMP-dependent pathway.
- cAMP is synthesised from ATP by adenylyl cyclase located on the inner side of the plasma membrane. *Adenylyl cyclase* is activated by a range of signaling molecules through the activation of *adenylyl cyclase* stimulatory G (Gs)-protein-coupled receptors and inhibited by agonists of *adenylyl cyclase* inhibitory G (Gi)-protein-coupled receptors. Liver *adenylyl cyclase* responds more strongly to glucagon, and muscle adenylyl cyclase responds more strongly to adrenaline.
- cAMP decomposition into AMP is catalyzed by the enzyme *phosphodiesterase*.
- Function: cAMP is a second messenger, used for intracellular signal transduction, such as transferring the effects of hormones like glucagon and adrenaline, which cannot pass through the cell membrane. It is involved in the activation of *protein kinases* and regulates the effects of adrenaline and glucagon. It also regulates the passage of Ca^{2+} through ion channels. cAMP and its associated kinases function in several biochemical processes, including the regulation of glycogen, sugar, and lipid metabolism by activating protein kinase.

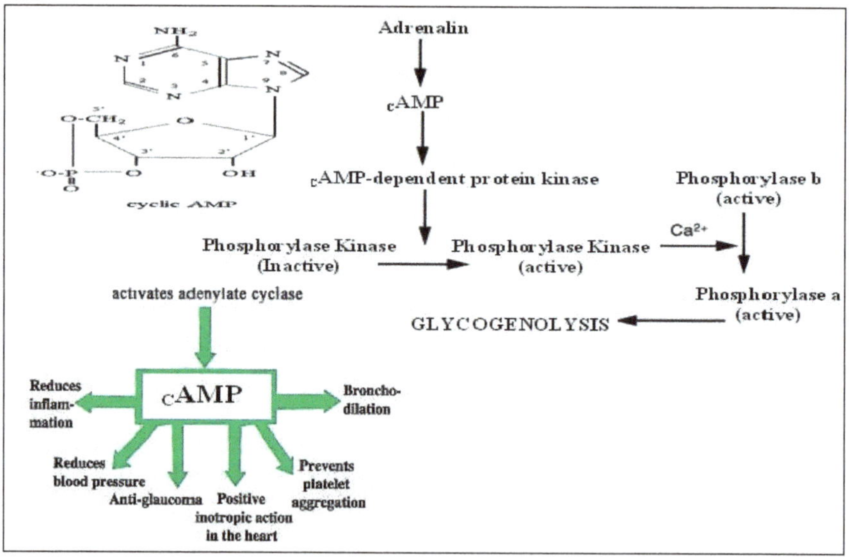

Guanosine Triphosphate (GTP)

- Guanosine-5'-triphosphate (GTP) is a purine nucleoside triphosphate. It can act as a substrate for both the synthesis of RNA during the transcription process and of DNA during DNA replication.

- It also has the role of a source of energy or an activator of substrates in metabolic reactions, like that of ATP, but more specific. It is used as a source of energy for protein synthesis and gluconeogenesis.
- GTP is essential to signal transduction, in particular with G-proteins, in second-messenger mechanisms where it is converted to Guanosine diphosphate (GDP) through the action of GTPases.
- Uses:
  - Energy transfer: GTP is involved in energy transfer within the cell. For instance, a GTP molecule is generated by one of the enzymes in the citric acid cycle. This is tantamount to the generation of one molecule of ATP, since GTP is readily converted to ATP with nucleoside-diphosphate kinase (NDK).
  - Genetic translation: During the elongation stage of translation, GTP is used as an energy source for the binding of a new amino- bound tRNA to the A site of the ribosome.
  - Mitochondrial function: The translocation of proteins into the mitochondrial matrix involves the interactions of both GTP and ATP.
- Synthesis of AMP and GMP from IMP.

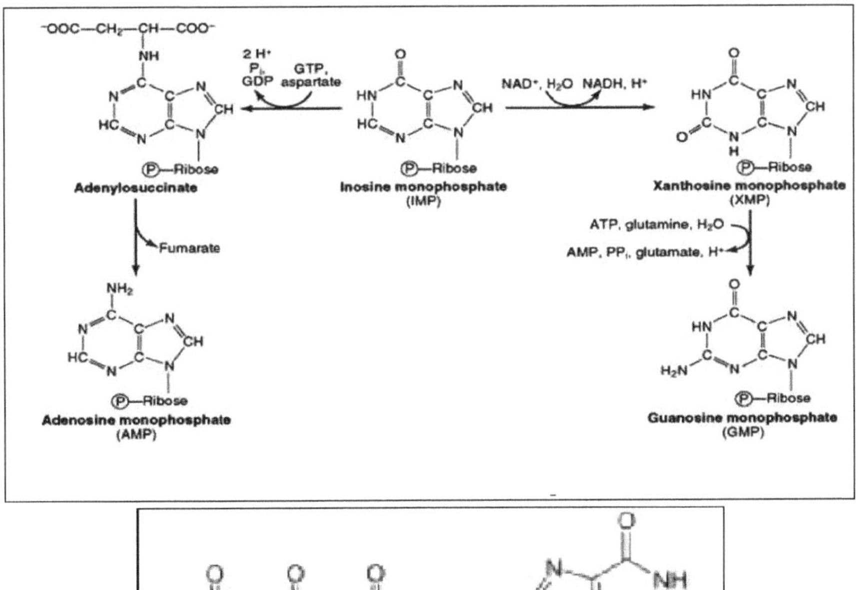

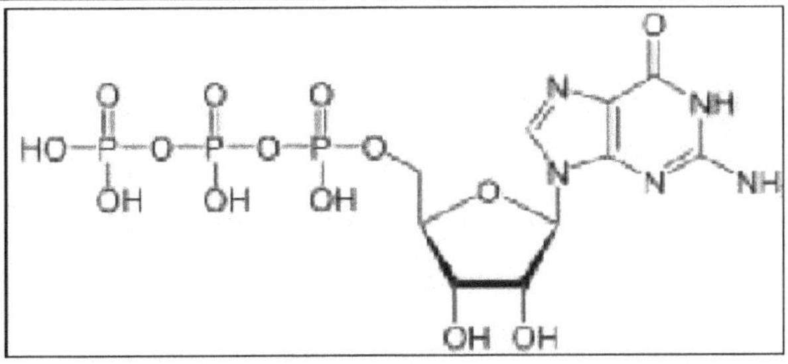

Guanosine Triphosphate.

Cyclic Guanosine Monophosphate

- Cyclic guanosine monophosphate (cGMP) is a cyclic nucleotide derived from guanosine triphosphate (GTP).
- cGMP acts as a second messenger much like cyclic AMP. Its most likely mechanism of action is activation of intracellular protein kinases in response to the binding of membrane-impermeable peptide hormones to the external cell surface.
- Synthesis: Guanylate cyclase (GC) catalyzes cGMP synthesis. This enzyme converts GTP to cGMP.

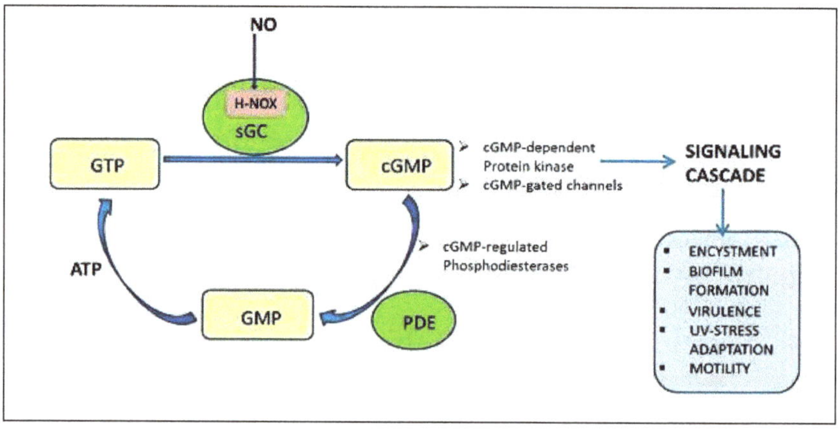

Figure shows schematic representation of synthesis, degradation, and function of cGMP. The three targets of cGMP molecules are (i) cGMP dependent protein kinases, (ii) cGMP gated ion channels and (iii) cGMP-dependent phosphodiesterases. While phosphodiesterases are involved in the degradation of cGMP to GMP, the protein kinases and activation of ion channels are subsequently involved in various bacterial signaling pathways. GTP: Guanosine 5′-triphosphate; sGC: soluble guanylate cyclase; NO: nitric oxide; H-NOX: Heme-Nitric oxide/Oxygen domain; cGMP: cyclic guanosine 3′,5′-monophosphate; PDE: Phosphodiesterase; GMP: guanosine 3′,5′- monophosphate; ATP: adenosine 5′-triphosphate.

Effects

- cGMP is a common regulator of ion channel conductance, glycogenolysis, and cellular apoptosis. It also relaxes smooth muscle tissues. In blood vessels, relaxation of vascular smooth muscles leads to vasodilation and increased blood flow.
- cGMP is a secondary messenger in phototransduction in the eye.
- cGMP is involved in the regulation of some protein-dependent kinases.

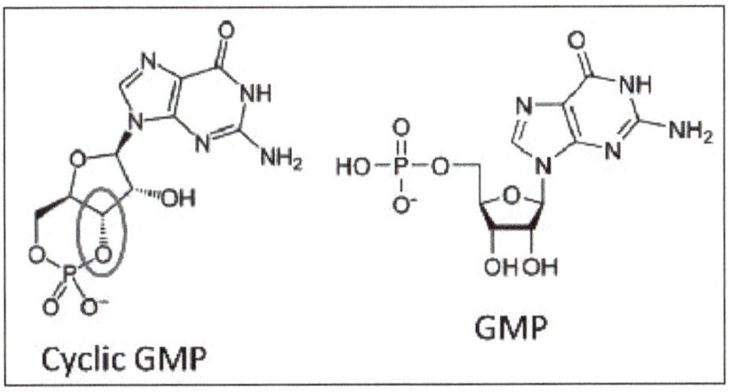

Glucose Storage

Carbohydrates or saccharides are the most abundant biological molecules. The basic carbohydrate units are called monosaccharides. There are numerous different types of monosaccharides, which differ in their number of carbon atoms and in the arrangement of the H and O atoms attached to the carbons. Furthermore, monosaccharides can be strung together in almost limitless ways to form polysaccharides. Sucrose which is present in common table sugar is broken down into its monomeric unit i.e. glucose. Glucose is the sugar which gives the energy that our body needs. When we intake carbohydrate rich food, it is broken down into its smallest molecule i.e. glucose. From the intestinal tract it goes into the bloodstream which carries these molecules throughout the body. Under usual conditions, glucose is the brain's primary fuel. Since brain cells store very little glycogen, they require a steady supply of glucose from the blood. A blood glucose concentration of less than half the normal value of ≈ 5 mM results in brain dysfunction. Levels much below this result in coma, irreversible damage, and ultimately death.

Sucrose present in common table sugar.

Once an organism has taken food, the food is digested, and needed nutrients are sent through the bloodstream. When the organism has used all the nutrients it needs to maintain proper functioning, the remaining nutrients are excreted or stored. Excess glucose is stored in the liver as the large compound called glycogen. Glycogen, the storage polysaccharide of animals, is present in all cells but is most prevalent in skeletal muscle and in liver, where it occurs as cytoplasmic granules. Glycogen has a highly branched structure which has many non-reducing ends so it permits the rapid mobilization of glucose when there is a metabolic need.

Food Sources of Glucose

Glucose is the most simple sugar of carbohydrates. When more complex carbohydrates such as polysaccharides and disaccharides are broken down in the stomach, they break down into the monosaccharide glucose. Carbohydrates serve as the primary energy source for working muscles, help brain and nervous system functioning and help the body use fat more efficiently.

Carbohydrates are present in most of the food we intake such as bread, fruits, vegetables, meats and so on. Any food that contains sugar has carbohydrates. And, most foods are converted to sugars when they are digested.

Food sources of Glucose or Carbohydrates.

Function of Glucose

Two most important function of glucose are:

- The primary function of glucose is to provide energy for physiological processes such as respiration, muscle contraction and relaxation, heart rhythm and the regulation of body temperature.
- The energy needed by the brain, neurons and developing red blood cells can only be acquire from glucose.

Glycolysis

Complete pathway of glucose breakdown is named as glycolysis. It is probably the most completely understood biochemical pathway which is a sequence of 10 enzymatic reactions as depicted in figure. It plays a key role in energy metabolism by providing a significant portion of the free energy used by most organisms and by preparing glucose and other compounds for further oxidative degradation.

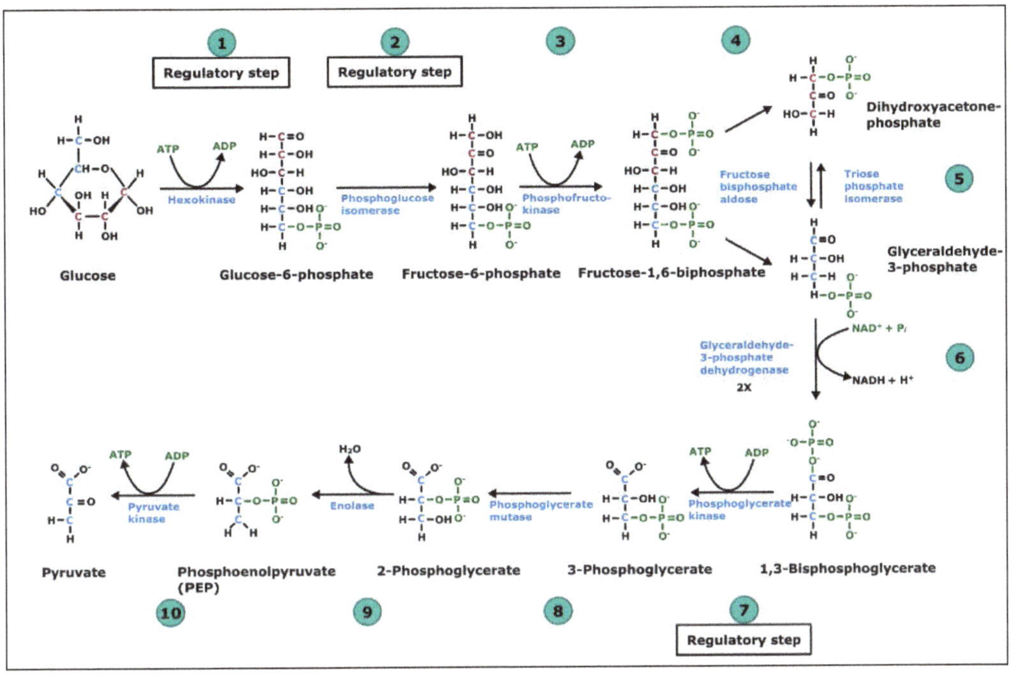

Glycolytic pathway.

Glucose usually appears in the blood as a result of the breakdown of polysaccharides (eg. Liver glycogen or dietary starch and glycogen) or from its synthesis from noncarbohydrate precursors (gluconeogenesis). Glucose enters most cells by a specific carrier that transports it from the exterior of the cell into the cytosol.

Glycolysis converts glucose to two C3 units (pyruvate). The free energy released in the process is harvested to synthesize ATP from ADP and Pi. Note that ATP is used early in the pathway to synthesize phosphorylated compounds but is alter resynthesized twice over. Glycolysis can therefore be divided into two stages:

- Stage I: Energy investment- In this preparatory stage, the hexose glucose is phosphorylated and cleaved to yield two molecules of triose glyceraldehydes-3-phosphate. This process consumes 2 ATP.
- Stage II: Energy recovery- The two molecules of glyceraldehydes-3-phosphate are converted to pyruvate, with generation of 4 ATP. Glycolysis therefore has a net profit of 2 ATP per glucose.

The phosphoryl groups that are initially transferred from ATP to the hexose do not immediately result in 'high energy' compounds. However, subsequent enzymatic transformations convert these 'low energy' products to compounds with high phosphoryl group-transfer potentials, which are capable of phosphorylating ADP to form ATP.

The overall reaction in the glycolysis goes like:

Glucose + $2P_i$ + 2ADP + $2NAD^+$ → 2 pyruvate + 2ATP + $2NADH^+$ + $2H^+$ + $2H_2O$

Storage of Glucose

Once carbohydrates are absorbed from food, they are carried to the liver where its processing occurs. Some glucose is sent to the bloodstream for its availability to various tissues while the rest is stored for later requirement of energy. It is stored in two types as muscle glycogen and as liver glycogen.

Muscle

Muscle stores approximately 1200 kcal or 5020 kJ amount of glycogen. This glycogen is readily converted into glucose 6-phosphate for entry into glycolysis. Muscle cannot export glucose because it lacks glucose -6-phosphatase. Furthermore, although muscle can synthesize glycogen from glucose, it does not participate in gluconeogenesis because it lacks the required enzymatic machinery. Consequently, muscle carbohydrate metabolism serves only muscle. In actively contracting skeletal muscle, much of the pyruvate formed is reduced to lactate, some of which flows to the liver, where it is con-verted into glucose.

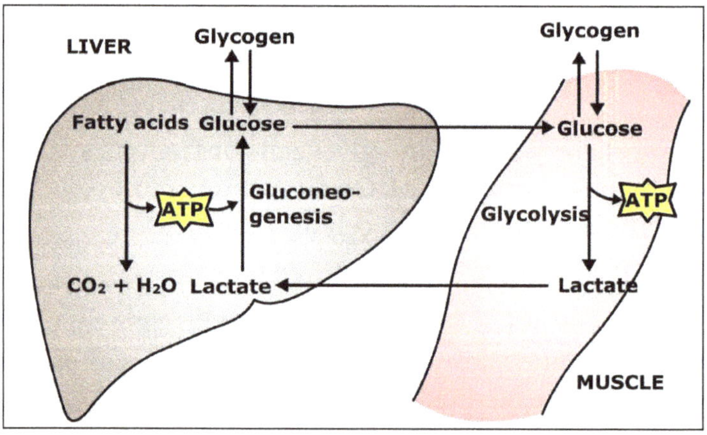

Metabolic interchanges between muscle and liver.

Liver

The liver maintains the proper levels of circulating fuels for use by various tissues such as brain and muscles. One of the major functions of liver is to act as a blood glucose 'buffer'. It does so by taking up and releasing glucose in response to hormones and to the concentration of glucose itself. After a carbohydrate containing meal, when the blood glucose concentration reaches ≈ 6mM, the liver takes up glucose by converting it to glucose-6-phosphate. The liver can produce glucose for release into the blood by breaking down its store of glycogen and by carrying out gluconeogenesis.

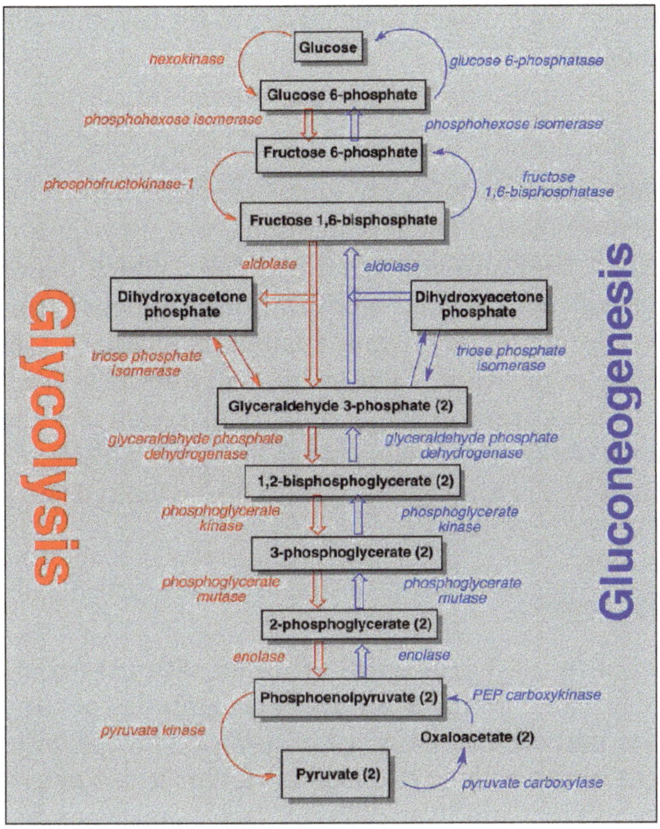

Gluconeogenesis.

Storage of Glucose in Plants

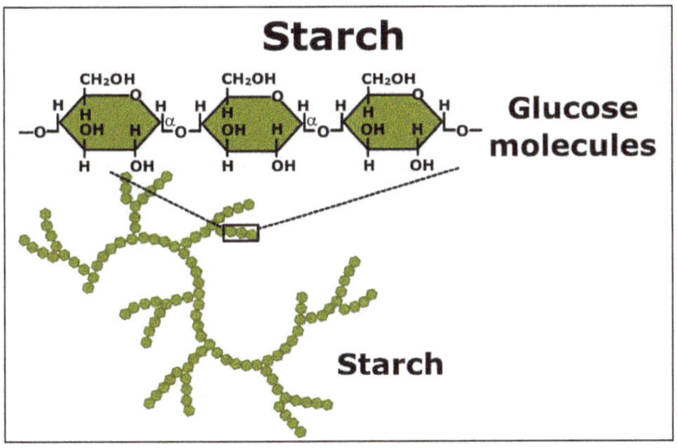

Structure of starch.

The storage form of glucose in plants is starch. Starch is a polysaccharide of the monomer glucose as can be seen in the figure. The leaves of a plant make sugar during the process of photosynthesis using chlorophyll present in them. Photosynthesis occurs in light (photo = light) when the sun is shining as the energy from the sunlight is used to make energy for the plant. So, when plants are making sugar, they store some of it as starch. When the simple sugars need to be retrieved for use, the starch is broken down into its smaller components.

Glycogen

Glycogen (in animals, fungi and bacteria) and starch (in plants) can function to stockpile glucose for later metabolic use. In animals, a constant supply of glucose is essential for tissues such as the brain and red blood cells, which depend almost entirely on glucose as an energy source. The mobilization of glucose from glycogen stores, primarily in the liver, provides a constant supply of glucose ($\approx$ 5 mM in blood) to all tissues. When glucose is plentiful, such as immediately after a meal, glycogen synthesis accelerates. Yet the liver's capacity to store glycogen is sufficient to supply the brain with glucose for about half a day. Under fasting conditions, most of the body's glucose needs are met by gluconeogenesis from non-arbohydrate precursors such as amino acids.

Structure of Glycogen

Glycogen is a polymer of α(1→4)-linked D-glucose with α(1→6)-linked branches every 8-14 residues. Glycogen occurs as intracellular granules of 100 to 400 Å diameter spheroidal molecules that each contain up to 120,000 glucose units. The granules are especially prominent in the cells that make the greatest use of glycogen: muscle (up to 1-2% glycogen by weight) and liver cells (up to 10% glycogen by weight). Glycogen granules also contain the enzymes that catalyze glycogen synthesis and degradation as well as many of the proteins that regulate these processes.

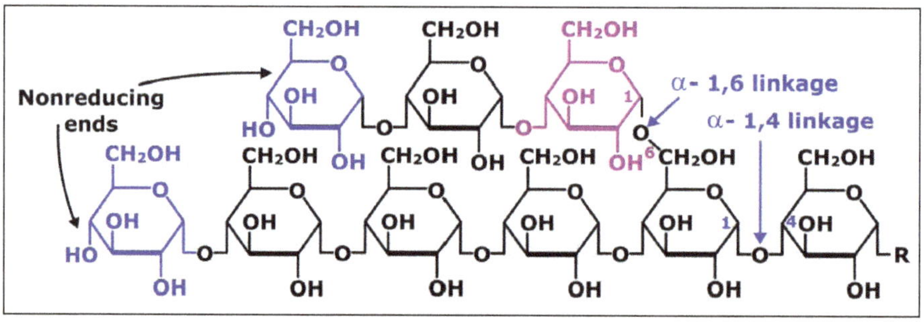

Structure of Glycogen.

In this structure of two outer branches of a glycogen molecule, the residues at the nonreducing ends are shown in red and residue that starts a branch is shown in green. The rest of the glycogen molecule is represented by R.

Glucose units are mobilized by their sequential removal from the nonreducing ends of glycogen (the ends lacking a C1-OH group). Whereas glycogen has only one reducing end, there is a nonreducing end on every branch. Glycogen's highly branched structure therefore permits rapid glucose mobilization through the simultaneous release of the glucose units at the end of every branch.

Glycogen Metabolism

Glycogen degradation and synthesis are relatively simple biochemical processes. First let us see how degradation of glycogen takes place. Glycogen degradation consists of three steps: (1) the release of glucose 1-phosphate from glycogen, (2) the remodeling of the glycogen substrate to permit further degradation, and (3) the conversion of glucose 1- phosphate into glucose 6-phosphate for further metabolism.

Glucose 6-phosphate is produced by the phosphorylation of free glucose, by glycogen degradation, and by gluconeogenesis. It is also a precursor for glycogen synthesis and the pentose phosphate pathway. The liver can hydrolyze glucose 6-phosphate to glucose. Glucose is metabolized by glycolyis to pyruvate, which can be further broken down to acetyl-CoA for oxidation by the citric acid cycle. Lactate and amino acids, which are reversibly converted to pyruvate, are precursors for gluconeogenesis.

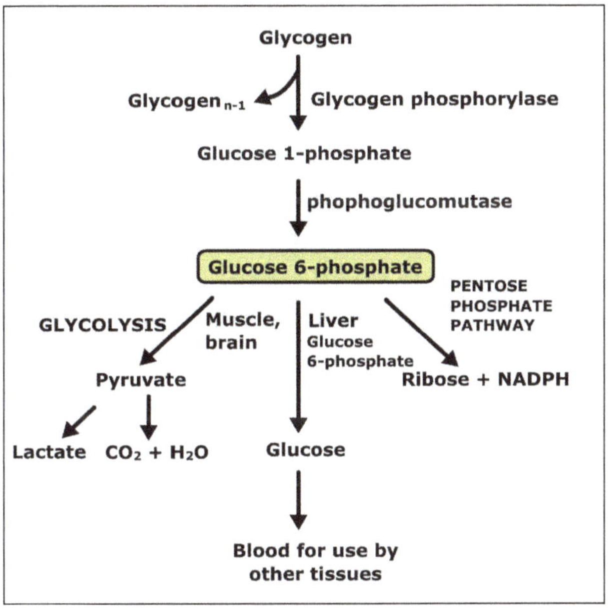

Fates of Glucose 6-Phosphate.

Chlorophyll

Chlorophyll is the molecule that traps this 'most elusive of all powers' - and is called a photoreceptor. It is found in the chloroplasts of green plants, and is what makes green plants, green. The basic structure of a chlorophyll molecule is a porphyrin ring, co-ordinated to a central atom. Initially, it was assumed that chlorophyll was a single compound but in 1864 Stokes showed by spectroscopy that chlorophyll was a mixture. If dried leaves are powdered and digested with ethanol, after concentration of the solvent, 'crystalline' chlorophyll is obtained, but if ether or aqueous acetone is used instead of ethanol, the product is 'amorphous' chlorophyll.

Chlorophyll was a mixture of two compounds, chlorophylla and chlorophyll-b:

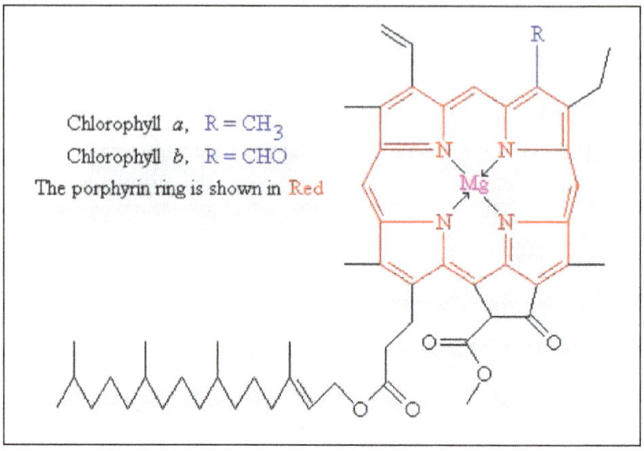

Chlorophyll-a (C55H72MgN4O5, mol. wt.: 893.49). The methyl group is replaced by an aldehyde in chlorophyll-b (C55H70MgN4O6, mol. wt.: 906.51).

The two components were separated by shaking a light petroleum solution of chlorophyll with aqueous methanol: chlorophyll-a remains in the light petroleum but chlorophyll-b is transferred into the aqueous methanol. Chlorophyll-a is a bluish-black solid and chlorophyll-b is a dark green solid, both giving a green solution in organic solutions. In natural chlorophyll there is a ratio of 3 to 1 (of a to b) of the two components. Both of these two chlorophylls are very effective photoreceptors because they contain a network of alternating single and double bonds, and the orbitals can delocalise stabilising the structure. Such delocalised polyenes have very strong absorption bands in the visible regions of the spectrum, allowing the plant to absorb the energy from sunlight.

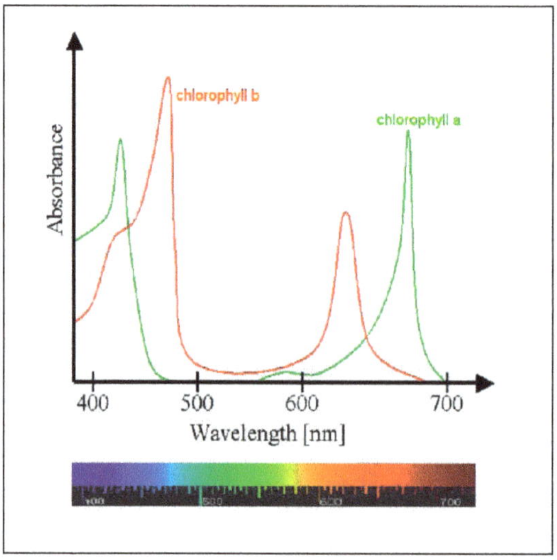

Absorption spectra of Chlorophyll a and b.

The different side groups in the 2 chlorophylls 'tune' the absorption spectrum to slightly different wavelengths, so that light that is not significantly absorbed by chlorophyll a, at, say, 460nm, will instead be captured by chlorophyll b, which absorbs strongly at that wavelength. Thus these two kinds of chlorophyll complement each other in absorbing sunlight. The absorption spectrum of chlorophyll shows that it absorbs strongly in the red and blue-violet regions of the visible spectrum. Because it absorbs red and blue-violet light, the light it reflects and transmits appears green.

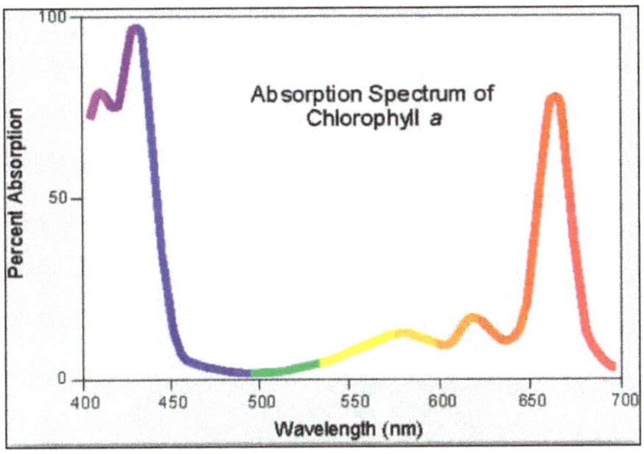

The uv/visible adsorption spectrum for chlorophyll.

Chlorophyll absorbs so strongly that it can mask other less intense colours. Some of these more delicate colours (from molecules such as carotene and quercetin) are revealed when the chlorophyll molecule decays in the autumn, and the woodlands turn red, orange, and golden brown. Chlorophyll can also be damaged when vegetation is cooked, since the central Mg atom is replaced by hydrogen ions. This affects the energy levels within the molecule, causing its absorbance spectrum to alter. Thus cooked leaves change colour - often becoming a paler, insipid yellowy green.

Chlorophyll in leaves decays in the autumn (the green colour fades and is replaced by the oranges and reds of carotenoids).

The chlorophyll molecule is the active part that absorbs the sunlight, but just as with hemoglobin, in order to do its job (synthesising carbohydrates) it needs to be attached to the backbone of a very complicated protein. This protein may look haphazard in design, but it has exactly the correct structure to orient the chlorophyll molecules in the optimal position to enable them to react with nearby CO_2 and H_2O molecules in a very efficient manner. Several chlorophyll molecules are lurking inside this bacterial photoreceptor protein.

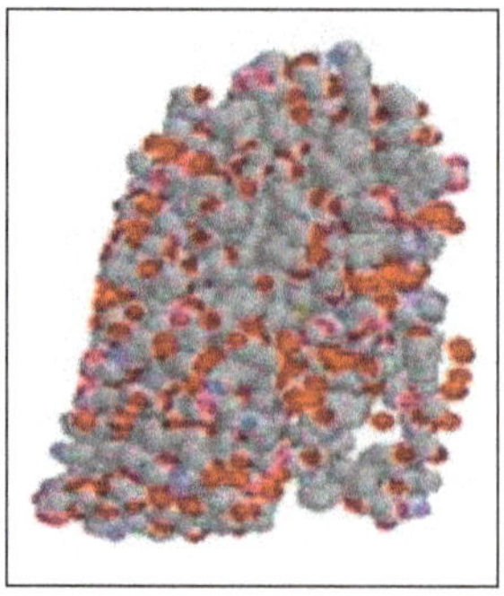

Bacterial photoreceptor protein.

Due to the green colour of chlorophyll, it has many uses as dyes and pigments. It is used in colouring soaps, oils, waxes and confectionary.

Green colour of chlorophyll.

Chlorophyll's most important use, however, is in nature, in photosynthesis. It is capable of channelling the energy of sunlight into chemical energy through the process of photosynthesis. In this process the energy absorbed by chlorophyll transforms carbon dioxide and water into carbohydrates and oxygen:

$$CO_2 + H_2O\ (CH_2 \rightarrow O) + O_2$$

The chemical energy stored by photosynthesis in carbohydrates drives biochemical reactions in nearly all living organisms. The actual chemical equation which takes place is the reaction between carbon dioxide and water, catalysed by sunlight, to produce glucose and a waste product, oxygen. The glucose sugar is either directly used as an energy source by the plant for metabolism or growth, or is polymerised to form starch, so it can be stored until needed. The waste oxygen is excreted into the atmosphere, where it is made use of by plants and animals for respiration.

$$6CO_2 + 6H_2O \xrightarrow[\text{Chlorophyll}]{\text{Sunlight}} \underset{\text{glucose}}{C_6H_{12}O_6} + 6O_2$$

In the photosynthetic reaction electrons are transferred from water to carbon dioxide i.e. carbon dioxide is reduced by water. Chlorophyll assists this transfer as when chlorophyll absorbs light energy, an electron in chlorophyll is excited from a lower energy state to a higher energy state. In this higher energy state, this electron is more readily transferred to another molecule. This starts a chain of electron-transfer steps, which ends with an electron being transferred to carbon dioxide. Meanwhile, the chlorophyll which gave up an electron can accept an electron from another molecule. This is the end of a process which starts with the removal of an electron from water. Thus, chlorophyll is at the centre of the photosynthetic oxidation-reduction reaction between carbon dioxide and water.

Other molecules with structures similar to that of chlorophyll play important roles in other biochemical electron-transfer (oxidation-reduction) reactions. Heme consists of a porphyrin similar to that in chlorophyll and an iron(II) ion in the center of the porphyrin. Heme is bright red. In the red blood cells of vertebrates, heme is bound to proteins forming hemoglobin. Hemoglobin combines with oxygen in the lungs, gills, or other respiratory surfaces and releases it in the tissues. In muscle cells, myoglobin, the name given to hemoglobin in muscles, stores oxygen as an electron source for energy-releasing oxidation-reduction reactions.

Another relative of chlorophyll is vitamin B_{12}. Vitamin B_{12} contains a cobalt ion at the center of the porphyrin. Like heme, vitamin B_{12} is bright red. It is essential to digestion and nutritional absorption in animals. The exact way it functions is not known. Because vitamin B12 is not produced by higher plants, a strictly vegetarian diet can lead to vitamin B_{12} deficiency. However, it is produced by molds and bacteria which grow on most foods.

The intense color of chlorophyll suggests that it may be useful as a commercial pigment. In fact, chlorophyll a is a green dye (Natural Green 3) used in soaps and cosmetics. Commercial pigments with structures similar to chlorophyll have been produced in a range of colors. Some of these have slightly modified porphyrins, such as having hydrogen atoms replaced with chlorine atoms. Others have different metal ions. For example, one bright blue pigment has a copper(I) ion at the center of the porphyrin and is used primarily in coloring fabrics.

Simple Reactions of Chlorophyll

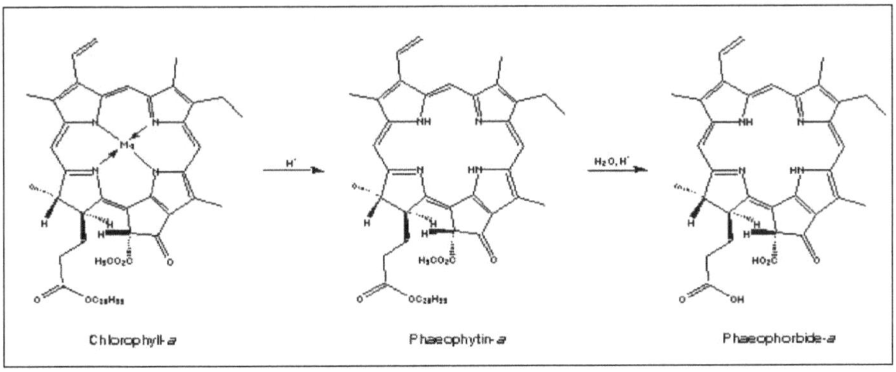

Overall reaction scheme for the hydrolysis of chlorophyll.

Treatment of cholorophyll-a with acid removes the magnesium ion replacing it with two hydrogen atoms giving an olive-brown solid, phaeophytin-a. Hydrolysis of this (reverse of esterification) splits off phytol and gives phaeophorbide-a. Similar compounds are obtained if chlorophyll-b is used.

Chlorophyll can also be reacted with a base which yields a series of phyllins, magnesium porphyrin compounds. Treatment of phyllins with acid gives porphyrins.

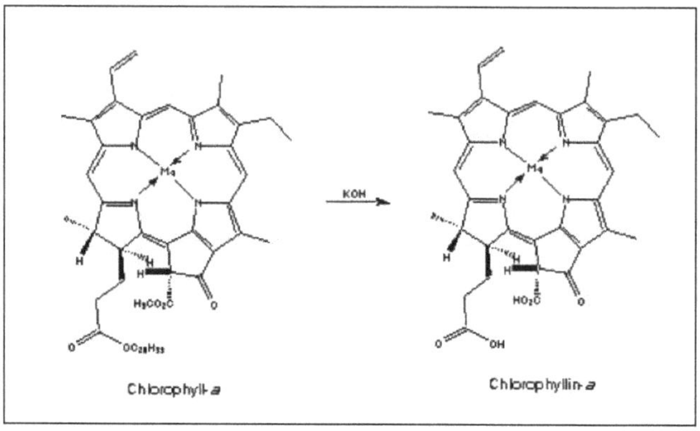

Overall scheme for the reaction of alkaline with chlorophyll.

Extraction of Chlorophyll from Plants

In plants chlorophyll is associated with specific proteins, for example, chlorophyll-a binding proteins are referred to as CP I, CP 47 and CP 43. With improving biochemical techniques for use on the membrane systems there has been an ever increasing success in the isolation and characterization of these proteins.

Initially, detergents are used to break down the membrane into fragments, and these fragments are further broken down by the use of different detergents. These detergents work by replacing the membrane lipids which surround integral membrane proteins. The resulting particles are separated by polyacrylamide gel electrophoresis (a standard biochemical method) in the presence of sufficient detergent to keep them 'solubilised'. The activity and polypeptide composition can then be assayed as the particle is purified. The detergents work by substituting lipids at different spots in the membrane, this is also affected by the concentration of the detergent. One such detergent that is very commonly used is SDS-PAGE (sodium dodecyl sulfate-polyacrylamide). This is generally used as it has several advantages over other detergents: the separation can be carried out fairly rapidly and it also gives a good overall picture of the distribution of chlorophyll.

Health Benefits of Chlorophyll

One of the best health benefits we gain from following a raw, living foods diet are the numerous health benefits from chlorophyll. One of the most interesting aspects of chlorophyll is how closely it resembles our red blood cells, known as hemoglobin, the pigment that gives our blood its red color as well as oxygen-carrying capacity. The hemoglobin of the red blood cell and the chlorophyll of the plant are virtually identical in molecular structure, with the only difference being the center atom. In hemoglobin it's iron and in chlorophyll it's magnesium. Chlorophyll is even often referred to as 'the green blood of plants' – obviously for good reason.

Blood Cleanser, Blood Builder and Oxygen Booster

This extreme similarity is what makes chlorophyll so beneficial to our health because it's such a powerful blood cleanser and blood builder. Chlorophyll delivers a continuous energy transfusion into our bloodstream, replenishing and increasing red blood cell count. And since hemoglobin carries oxygen to our cells, increasing hemoglobin thereby increases the ability of red blood cells to carry oxygen and delivers us increased levels of oxygen – and this is a very good thing.

Disease-promoting anaerobic bacteria cannot live and thrive in the presence of oxygen. It's also known that disease can't thrive in an alkaline environment either. Lucky for us, chlorophyll is both a potent oxygenator and helps alkalize our blood; it thus helps our bodies fight off disease. All these benefits direct effects on strengthening our immune system and enhancing overall energy levels and wellbeing.

Detoxification

Healthy blood flow and an abundance of oxygen also help the body to cleanse itself of toxic impurities, lending to chlorophylls detoxification properties. Chlorophyll has also been shown to improve the detrimental effects of radiation, and has also been linked to cancer prevention. Chlorophyll binds with toxins like heavy metals and helps remove them from the body. Chlorophyll can also stimulate bowel movements and is a great aid in colon cleansing.

Overall Benefits of Chlorophyll

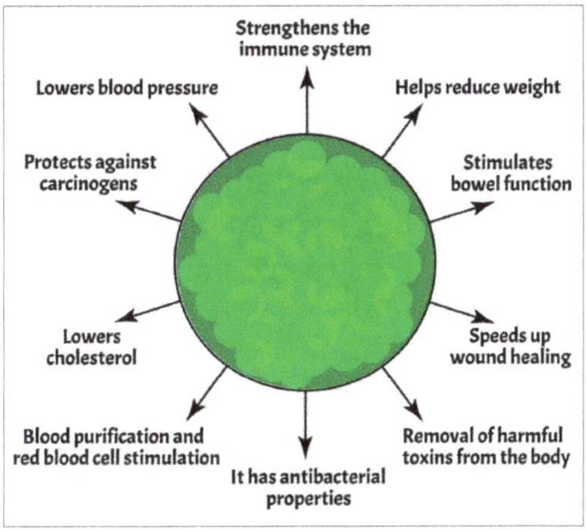

Benefits of Chlorophyll.

The range of health benefits to gain from chlorophyll is nothing short of miraculous. Chlorophyll is known to improve the health of the circulatory, digestive, immune and detoxification systems of the body – leading to many different and overlapping health benefits.

The numerous health benefits of chlorophyll:

- Cleanses and oxygenates and builds the blood.
- A powerful detoxification effect on the body.
- Rich in enzymes that promote quick rejuvenation of our cells.
- High in Amino acids.
- Extracts toxins form the liver and improves liver function.
- Regulator of calcium.
- Helps break addiction.
- Alkalizes the body.
- Anti-inflammatory.
- Wound-healing properties.
- Fight infection.
- Anti-oxidant – neutralize free radicals.
- Promotes healthy intestinal flora.
- Helps reverse protein-deficient anemia.
- Protection from cancer.
- Helps skin disorders.

Sources of Chlorophyll

Essentially, any green plant that you consume contains some level of chlorophyll as this is what gives it its green color. But some foods are higher in chlorophyll than others. Generally speaking, the darker the green color, the more chlorophyll, so dark leafy greens are a great source, especially dino kale and other kale varieties, swiss chard and darker greens like arugula. Adding herbs like parsley and cilantro are also great ways to spruce up your salads and increase chlorophyll content in your diet. Blue-green algae such as spirulina and chlorella are also especially high in chlorophyll, as are all sprouts. Wheatgrass has one of the highest sources of chlorophyll available. Juicing it on a regular basis can have powerful detoxification effects.

Sources of Chlorophyll.

Chlorophylls are the most abundant pigments in plants. Dark green, leafy vegetables like spinach are rich sources of natural chlorophylls. The chlorophyll contents of selected vegetables are presented in the table below.

Table: Chlorophyll content of selected raw vegetable.

Food	Serving	Chlorophyll (mg)
Spinach	1 cup	23.7
Parsley	½ cup	19.0
Cress, garden	1 cup	15.6
Green beans	1 cup	8.3
Arugula	1 cup	8.2
Leeks	1 cup	7.7
Endive	1 cup	5.2
Sugar peas	1 cup	4.8
Chinese cabbage	1 cup	4.1

Permissions

All chapters in this book are published with permission under the Creative Commons Attribution Share Alike License or equivalent. Every chapter published in this book has been scrutinized by our experts. Their significance has been extensively debated. The topics covered herein carry significant information for a comprehensive understanding. They may even be implemented as practical applications or may be referred to as a beginning point for further studies.

We would like to thank the editorial team for lending their expertise to make the book truly unique. They have played a crucial role in the development of this book. Without their invaluable contributions this book wouldn't have been possible. They have made vital efforts to compile up to date information on the varied aspects of this subject to make this book a valuable addition to the collection of many professionals and students.

This book was conceptualized with the vision of imparting up-to-date and integrated information in this field. To ensure the same, a matchless editorial board was set up. Every individual on the board went through rigorous rounds of assessment to prove their worth. After which they invested a large part of their time researching and compiling the most relevant data for our readers.

The editorial board has been involved in producing this book since its inception. They have spent rigorous hours researching and exploring the diverse topics which have resulted in the successful publishing of this book. They have passed on their knowledge of decades through this book. To expedite this challenging task, the publisher supported the team at every step. A small team of assistant editors was also appointed to further simplify the editing procedure and attain best results for the readers.

Apart from the editorial board, the designing team has also invested a significant amount of their time in understanding the subject and creating the most relevant covers. They scrutinized every image to scout for the most suitable representation of the subject and create an appropriate cover for the book.

The publishing team has been an ardent support to the editorial, designing and production team. Their endless efforts to recruit the best for this project, has resulted in the accomplishment of this book. They are a veteran in the field of academics and their pool of knowledge is as vast as their experience in printing. Their expertise and guidance has proved useful at every step. Their uncompromising quality standards have made this book an exceptional effort. Their encouragement from time to time has been an inspiration for everyone.

The publisher and the editorial board hope that this book will prove to be a valuable piece of knowledge for students, practitioners and scholars across the globe.

Index

CPSIA information can be obtained
at www.ICGtesting.com
Printed in the USA
LVHW060908030122
707669LV00002B/398

9 781639 874552